Lecture Notes in Networks and Systems

Volume 580

The series "Lecture Notes in Networks and Systems" publishes the latest developments in Networks and Systems—quickly, informally and with high quality. Original research reported in proceedings and post-proceedings represents the core of LNNS.

Volumes published in LNNS embrace all aspects and subfields of, as well as new challenges in, Networks and Systems.

The series contains proceedings and edited volumes in systems and networks, spanning the areas of Cyber-Physical Systems, Autonomous Systems, Sensor Networks, Control Systems, Energy Systems, Automotive Systems, Biological Systems, Vehicular Networking and Connected Vehicles, Aerospace Systems, Automation, Manufacturing, Smart Grids, Nonlinear Systems, Power Systems, Robotics, Social Systems, Economic Systems and other. Of particular value to both the contributors and the readership are the short publication timeframe and the world-wide distribution and exposure which enable both a wide and rapid dissemination of research output.

The series covers the theory, applications, and perspectives on the state of the art and future developments relevant to systems and networks, decision making, control, complex processes and related areas, as embedded in the fields of interdisciplinary and applied sciences, engineering, computer science, physics, economics, social, and life sciences, as well as the paradigms and methodologies behind them.

Indexed by SCOPUS, INSPEC, WTI Frankfurt eG, zbMATH, SCImago.

All books published in the series are submitted for consideration in Web of Science.

For proposals from Asia please contact Aninda Bose (aninda.bose@springer.com).

More information about this series at https://link.springer.com/bookseries/15179

Marco Temperini · Vittorio Scarano ·
Ivana Marenzi · Milos Kravcik ·
Elvira Popescu · Rosa Lanzilotti ·
Rosella Gennari · Fernando De La Prieta ·
Tania Di Mascio · Pierpaolo Vittorini
Editors

Methodologies and Intelligent Systems for Technology Enhanced Learning, 12th International Conference

Springer

Editors
Marco Temperini
Sapienza University of Rome
Rome, Roma, Italy

Ivana Marenzi
Leibniz University of Hanover
Hanover, Germany

Elvira Popescu
Computer and Information Technology Department
University of Craiova
Craiova, Romania

Rosella Gennari
Faculty of Computer Science
Free University of Bozen-Bolzano
Bolzano, Italy

Tania Di Mascio
DISIM
University of L'Aquila
L'Aquila, L'Aquila, Italy

Vittorio Scarano
University of Salerno
Montoro - Avellino, Italy

Milos Kravcik
DFKI GmbH
Berlin, Germany

Rosa Lanzilotti
University of Bari Aldo Moro
Bari, Italy

Fernando De La Prieta
University of Salamanca
Salamanca, Spain

Pierpaolo Vittorini
University of L'Aquila
L'Aquila, Italy

ISSN 2367-3370 ISSN 2367-3389 (electronic)
Lecture Notes in Networks and Systems
ISBN 978-3-031-20616-0 ISBN 978-3-031-20617-7 (eBook)
https://doi.org/10.1007/978-3-031-20617-7

This Springer imprint is published by the registered company Springer Nature Switzerland AG
The registered company address is: Gewerbestrasse 11, 6330 Cham, Switzerland

Preface

Education is the cornerstone of any society, and it serves as one of the foundations for many of its social values and characteristics. State-of-the-art and novel methodologies and technologies allow researchers, designers, and domain experts to pursue Technology Enhanced Learning (TEL) solutions targeting not only cognitive processes but also motivational, personality, or emotional factors. Nowadays, we can identify two main legs, providing necessary and complementary strengths to a TEL-oriented design process: Appropriate technologies should be applied, and appropriate methods should guide such application. Technologies in TEL can deliver smart, personalized, tailored, and motivating learning solutions. Methods come from different fields, such as psychology, medicine, computer science, and from diverse communities, where collaboration and co-working are used, such as maker communities and participatory design communities. In addition, learning analytics can help manage big data to enhance learning opportunities for learners and educators alike, for instance, by supporting self-regulated learning or adaptation of the learning material.

As to these topics, the annual appointment of MIS4TEL established itself as a consolidated fertile forum where scholars and professionals from the international community, with a broad range of expertise in the TEL field, share results and compare experiences. The call for papers of the 12th edition of the conference welcomed novel research in TEL and expanded on the topics of the previous editions: It solicited work from new research fields, ranging from artificial intelligence and agent-based systems to robotics, virtual reality, Internet of things, and wearable solutions, among others, concerning methods and technological opportunities, and how they serve to create novel approaches to TEL, innovative TEL solutions, and valuable TEL experiences.

The result of the call for papers is that both the main tracks of MIS4TEL 2022 and its three related workshops such as Artificial Intelligence for Education (Ai4Ed), Technology Enhanced Learning in Nursing Education (NURSING), and Reflections and Dialogues around Smart Technology (ResiSTo) contribute to novel research in TEL and expand on the topics of the previous editions. This volume presents the papers that were accepted for the main track of MIS4TEL 2022.

All papers underwent a peer-review selection: Each paper was assessed by three different reviewers, from an international panel composed of about 65 members from 21 countries. From a total of 33 articles, the program of MIS4TEL 2022 counts 23 contributions, 12 full papers, and 11 short papers from diverse countries.

We would like to thank all the contributing authors, the members of the Program Committee, the reviewers, the sponsors, and the Organizing Committee for their hard and highly valuable work. Thanks for your help—MIS4TEL 2022 would not exist without your contribution.

Marco Temperini
Vittorio Scarano
Ivana Marenzi
Milos Kravcik
Elvira Popescu
Rosa Lanzillotti
Rosella Gennari
Fernando De La Prieta
Tania Di Mascio
Pierpaolo Vittorini

Organization of MIS4TEL 2022

http://www.mis4tel-conference.net/

General Chair

Marco Temperini	Sapienza University of Rome, Italy

Technical Program Chairs

Vittorio Scarano	University of Salerno, Italy
Ivana Marenzi	L3S Research Center, Leibniz Universität Hannover, Germany

Paper Chairs

Milos Kravcik	German Research Center for Artificial Intelligence, Germany
Elvira Popescu	University of Craiova, România
Rosa Lanzillotti	University of Bari, Italy

Workshop Chairs

Zuzana Kubincova	Comenius University in Bratislava, Slovakia
Alessandra Melonio	Ca' Foscari University of Venice, Italy

Publicity Chairs

Federica Caruso	University of L'Aquila, Italy
Agnese Addone	University of Salerno, Italy
John Jairo Páez Rodríguez	University Francisco José de Caldas of Bogotà, Colombia

Steering Committee Representatives

Rosella Gennari	Free University of Bozen-Bolzano, Italy
Fernando De la Prieta	University of Salamanca, Spain

Local Organizing Committee

Pierpaolo Vittorini (Co-chair)	University of L'aquila, Italy
Tania Di Mascio (Co-chair)	University of L'aquila, Italy
Federica Caruso	University of L'Aquila, Italy
Anna Maria Angelone	University of L'Aquila, Italy

Organizing Committee

Juan M. Corchado Rodríguez	University of Salamanca, Spain and AIR Institute, Spain
Fernando De la Prieta	University of Salamanca, Spain
Sara Rodríguez González	University of Salamanca, Spain
Javier Prieto Tejedor	University of Salamanca, Spain and AIR Institute, Spain
Pablo Chamoso Santos	University of Salamanca, Spain
Liliana Durón	University of Salamanca, Spain
Belén Pérez Lancho	University of Salamanca, Spain
Ana Belén Gil González	University of Salamanca, Spain
Ana De Luis Reboredo	University of Salamanca, Spain
Angélica González Arrieta	University of Salamanca, Spain
Emilio S. Corchado Rodríguez	University of Salamanca, Spain
Alfonso González Briones	University of Salamanca, Spain
Yeray Mezquita Martín	University of Salamanca, Spain
Beatriz Bellido	University of Salamanca, Spain
María Alonso	University of Salamanca, Spain
Sergio Marquez	University of Salamanca, Spain
Marta Plaza Hernández	University of Salamanca, Spain
Guillermo Hernández González	AIR Institute, Spain
Ricardo S. Alonso Rincón	University of Salamanca, Spain
Raúl López	University of Salamanca, Spain
Sergio Alonso	University of Salamanca, Spain
Andrea Gil	University of Salamanca, Spain
Javier Parra	University of Salamanca, Spain

Program Committee

Marie Helene Abel	Université de technologie de Compiègne, UTC, France
Agnese Addone	University of Salerno, Department of Computer Science, Italy
Juan M. Alberola	Universitat Politècnica de València, Spain
Peter Bednar	University of Portsmouth, UK
Rosa Maria Bottino	ITD-CNR, Italy
Tharrenos Bratitsis	University of Western Macedonia, Greece
Nicola Capuano	Università degli Studi della Basilicata, Italy
Davide Carneiro	CIICESI/ESTG, Polytechnic Institute of Porto, Portugal
Federica Caruso	University of L'Aquila, Italy
Maiga Chang	Athabasca University, Canada
Vincenza Cofini	University of L'Aquila, Italy
Cesar A. Collazos	Universidad de Cauca, Colombia
Mihai Dascalu	Computer Science and Engineering Department of University Politehnica of Bucharest, Romania
Giovanni De Gasperis	DISIM, Università degli Studi dell'Aquila, Italy
Maria De Marsico	Sapienza Università di Roma, Italy
Damiano Distante	University of Rome Unitelma Sapienza, Italy
Dalila Duraes	Universidade do Minho, Portugal
Florentino Fdez-Riverola	University of Vigo, Spain
Florentino Fernández Manjón	Universidad Complutense de Madrid, Spain
Margarida Figueiredo	Universidade de Évora, Portugal
Blanka Frydrychova Klimova	University of Hradec Kralove, Czech Republic
Denis Gillet	École polytechnique fédérale de Lausanne, Switzerland
Carlo Giovannella	University of Tor Vergata, Italy
Jorge Gomez-Sanz	Universidad Complutense de Madrid, Spain
Sabine Graf	Athabasca University, Canada
Eelco Herder	Radboud Universiteit Nijmegen, Netherlands
Mirjana Ivanovic	University of Novi Sad, Faculty of Sciences, Department of Mathematics and Informatics, Serbia
Vicente Julian	Universitat Politècnica de València, Spain
Ralf Klamma	RWTH Aachen University, Germany
Rosa Lanzillotti	University of Bari, Italy
Luigi Laura	International Telematic University Uninettuno, Italy
Élise Lavoué	Université Claude Bernard Lyon 1, France
Carla Limongelli	Universita' Roma Tre, Italy

Andreas Lingnau	University of Applied Sciences Ruhr West, Germany
Matteo Lombardi	Griffith University, Australia
George Magoulas	Department of Computer Science and Information Systems at Birkbeck, University of London, UK
Anna Mavroudi	Norwegian University of Science and Technology, Norway
Alessandra Melonio	Free University of Bozen-Bolzano, Italy
Marcelo Milrad	Linnaeus University, Sweden
Minoru Nakayama	Information and Communications Engineering, Tokyo Institute of Technology, Japan
Wolfgang Nejdl	L3S and University of Hannover, Germany
Alessandro Pagano	University of Bari, Italy
Kyparissia Papanikolaou	School of Pedagogical & Technological Education, Greece
Elvira Popescu	University of Craiova, Romania
Francesca Pozzi	Università di Bologna, Italy
Kasper Rodil	Aalborg University, Denmark
Sara Rodríguez	University of Salamanca, Spain
Iván Rodríguez Conde	University of Arkansas at Little Rock, USA
Veronica Rossano	University of Bari, Italy
Guiseppe Sansonetti	Roma Tre University, Italy
Olga C. Santos	Universidad Nacional de Educación a Distancia, UNED, Spain
Juan M. Santos	University of Vigo, Spain
Flippo Sciarrone	Roma Tre University, Italy
Andrea Sterbini	Sterbini, Italy
Davide Taibi	Italian National Research Council, Italy
Laura Tarantino	Università del'Aquila, Italy
Henrique Vicente	Universidade de Évora, Portugal

MIS4TEL'22 - Organizers and Sponsors

Sponsors

Organizers

Support from National Associations

Contents

Blended Learning in the Foundational Design Studio

Karma Dabaghi(✉) and Silia Abou Arbid

Lebanese American University, Beirut, Lebanon
{karma.dabaghi,silia.abouarbid}@lau.edu.lb

Abstract. Based on our experience of teaching the first-year Foundation studio online in an architecture program, we highlight the benefits of Blended Learning that incorporates strategic scheduling of physical and virtual delivery modes. Our qualitative observations build on recent literature that examines the use of online elements in teaching the design studio. We observed that the more theoretical aspects appear to be effectively mastered online while the practical ones that require tactile abilities are not. We identify what kinds of interaction maximize the potential of Technology Enhanced Learning (TEL) in the Foundation studio.

Keywords: Blended Learning · In-person learning · Foundation studio

1 Technology-Enhanced Learning (TEL) in the Design Studio

Design studio teaching in higher education relies on individualized interaction with instructors and collaboration with peers within shared environments. As students progress in their studies, they become sensitive to the importance of discussing ideas in groups and generating solutions collaboratively. We draw our observations from teaching the foundational studio during challenging times, where remote learning unexpectedly replaced in-person attendance in the classroom.

Studio classes are the core of design education, which has a long history of being based on individual guidance and frequent feedback from instructors and student peers [2–4]. Traditionally, students benefit from the proximity of the instructor and peers to interact in solving design problems. Informally, collective learning takes place through hours of open exchange on design ideas and proposals. Formally, feedback is given through the *desk crit,* or desk critique, where the student and the instructor review the work together, and the *pin-up,* where students exhibit their work to the entire class for review by the instructors. Peers participate in this discussion. At the midterm and final project review, a jury panel assesses the student work, gives feedback, and issues a formal evaluation [6].

Given this starting point, physical presence is necessary, and it is not surprising that design education has been slow to embrace TEL [3, 5, 7, 9]. The COVID-19 pandemic, however, jump-started opportunities to test different approaches to TEL in the design studio [4, 11]. New digital production technologies and easily accessible software and hardware (on phone and PC) together with innovative learning management

M. Temperini et al. (Eds.): MIS4TEL 2022, LNNS 580, pp. 1–6, 2023.
https://doi.org/10.1007/978-3-031-20617-7_1

systems (LMS) (such as Moodle or Blackboard) and videoconferencing platforms (such as WebEx or Zoom), facilitated experimentation.

TEL is impactful in design education through making formative feedback (desk crit, pin up, mid-term review, or final review) accessible to everyone. This keeps the instructor from correcting the same mistake at every desk, and gives the chance for students to benefit from each other [10, 11]. But there remains a persistent belief that regardless of student characteristics or preferences, a design studio that implements TEL will not ultimately succeed if it does not ensure what Crowther refers to as, "a reliable tactile learning experience as form of intrinsic feedback such as can be experienced in laboratory or workshop modes of delivery" [3].

2 Architectural Teaching at the First-Year Foundation Level

The Foundation Program curriculum introduces students to the fundamentals of visual perception and ensures that they develop a heightened critical thinking capacity when addressing design queries. Interpretational skills are enhanced with the use of conventional and new plastic media such as drawing and model fabrication at the woodshop.

Formal Tectonics is the first-year spring studio that introduces basic architectural concepts and explores the relationship between spatial constructs and human inhabitability through critical investigations of form, perception, and the human body. Projects mediate between the conceptual and the functional dimensions of spatial manifestations and address the complexities of theoretical discourse, context, nature and a site's history and prototypical conditions. The projects lead to carefully hand-drafted drawings and well-articulated models manually executed at the woodshop to engage design-thinking through *making*.

Formal Tectonics focuses on the elaboration of an *architectone*, a spatial object that is representative of potentially inhabitable space. The studio is open to a variety of teaching approaches that include reflective and deductive reasoning, sensorial investigations into the tectonics of materials, as well as the value of knowledge discovery through drawing. The studio requires an active engagement of thinking and making, and its outcome leads to well-articulated kinetic structures that materialize an engaging relationship between man, form, and space.

A focus on theory such as space poetics and phenomenology [1, 8] as it relates to the act of making is not unusual at the higher levels of architectural study [5]. Our approach, however, relies on theoretical reflection at the level of the Foundation studio to enhance the mind's ability to be resourceful, individual, and creative when approaching the acts of drawing and making informed by rigorous research on the history and theory of the creative act.

3 Case Study

This paper is based on the authors' observations of one instructor teaching two sections of the Foundation studio *Formal Tectonics,* over three semesters under different conditions without a teaching assistant. The first condition was completely in-person with 15 students per section. The two blended types are specified in Table 1.

Table 1. Identified types of Blended Learning.

Spring 2020	
Type 1 Blended Learning 15 students per section	40% taught fully in-person until the midterm 60% taught fully online in full lockdown until the end of the term
Spring 2021	
Type 2 Blended Learning 10 students per section	60% taught fully online until the withdraw date 35% taught in blended mode with intermittent in-person attendance 5% in-person follow-up after classes ended (final exam period)

3.1 Expected Studio Outcomes

In this studio, students are expected to create both 2D and 3D work. The 2D analytical construct is a technical drawing that includes a textural rendering component. Students execute these drawings with technical precision that yields a complex output. It is done with the technical drawing tools (a compass and a ruler) by applying a simple geometric theorem. Throughout the semester, students work on a series of exploratory drawings to achieve the expected level of technical skill.

The studio also focuses on mastering manual craftsmanship and using the woodshop as a laboratory. Students spend considerable time in the woodshop to sharpen their production skills, perform elaborate studies in kinetics, and understand morphology. As examples of this production, we chose the best students who could regularly work independently with or without peer interaction despite the unusual learning and working conditions be it in pre- or COVID-19 contexts. In this way, we attempted to isolate the mode of instruction from other variables, including COVID-induced anxiety.

3.2 Analysis and Comparison of Design Studio Outcomes

Fig. 1. Pre-COVID-19. Student work by Soumar Al Kamand.

Pre-COVID-19 In-Person Learning. The student whose work is featured in Fig. 1 took the studio entirely in-person. His work fulfilled expectations as outstanding in both its interpretive dimension and its technical execution. The output featured sophisticated detailing in both the model with the use of various materials and the drawing component demonstrating a highly sensitive and precise rendering in graphite and colored pencils

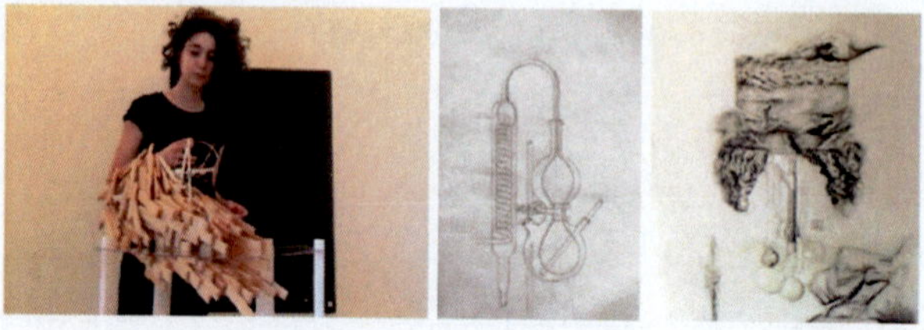

Fig. 2. Type **1 Blended Learning.** Student work by Tamara Nasr in the final presentation videoshort explaining her proposal for an architectural apparatus (spring 2020).

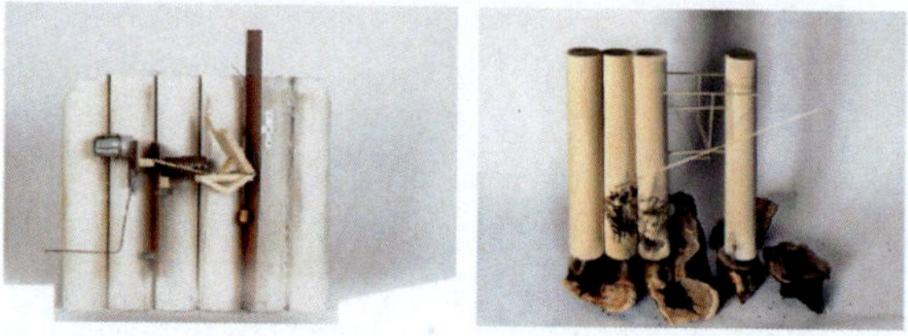

Fig. 3. Type **2 Blended Learning.** Student work by Jad Ghandour & Acyl Safa (spring 2021).

over an accurate construct of the insect's corporeality. This student benefited from more in-person contact with the instructor and interaction with his peers, which enhanced his learning and creative process. Because studio learning remains an interaction that is both social and intellectual, this outcome was expected. Open access to the woodshop resources allowed for the maturity of construction skills that were at the core of the studio.

Type 1 Blended Learning. The student whose work is featured in Fig. 2 was able to create a solid intellectual base for her project despite moving online. The analytical thinking, synthesis, and reflective reasoning led to a rigorous conceptual and aesthetic transcription of her research into 2D and 3D. Both drawing and model were successful in expressing a carefully researched and conceptualized proposal; however, both advanced technical rendition at the level of the 2D analytical construct and sophistication of the constructional details in the 3D model were underdeveloped. The translation of the student's research to 2D and 3D visuals was evident, yet because of the student's inability to access the school woodshop at a critical point until the end of the semester, the articulation of a well-developed *architectone* with mechanical components was impossible to achieve.

The online condition allowed the student to dedicate more time to her personal research and the development of her ideas. Another positive outcome of this condition was the jury that was held online. When juries could not be held in person, students were required to make an audio-visual presentation that showcased their work in two formats: a slideshow with voiceover and a videoshort with an oral presentation of their work similar to in-person final review panels. Most presentations featured elaborate editing techniques that the students had the curiosity and time to discover on their own.

Type 2 Blended Learning. The two students whose work is featured in Fig. 3 developed a complex concept for their project based on their research and their understanding of the theoretical component of the class. Acyl Safa's project (on the right) integrated Arduino circuit sensors (something she could learn on her own) to a set of rods held in tensegrity, that express her concept through elements within the model.

Jad Ghandour's model (on the left) was distinguished by a meticulously handcrafted telescopic oculus using metal sheets found at home. Ghandour researched landscapes that no longer exist and was able to successfully express his theoretical ideas through model making.

Both students expressed their appreciation of being online as it allowed for ample time to elaborate their personal quest for a strong concept, which included reading more and exploring research strategies new to novices in design studies. The instructor, having taught online in spring 2020, had developed an effective strategy for presenting well-formulated lecture content and keeping students engaged in the online environment, making sure that every student actively participated throughout the six-hour live streamed session. Students were continuously prompted to answer questions, take part in the feedback offered to peers, and launch debates on the thematic. These two *kinetic architectones* are highly conceptual but show lack of refinement due to the shop schedule being restricted in response to pandemic conditions. While formerly, students could spend unlimited time in the shop in the presence of a mentor, the restricted schedule during this term meant that peer communication was minimal, as was time that could be devoted to refinement of the models.

3.3 Observed Challenges and Advantages of Blended Learning

Online class lectures (WebEx) were conceived to be visually rich and were prepared ahead of time by the instructor to promote increased interest and levels of understanding. The teacher opted not to upload these lectures online to ensure maximum live engagement on behalf of the students and provide an environment that mimics in-person attendance in the classroom. Students were consistently prompted to give their feedback on the projects of their peers to replicate some aspects of studio culture.

Tutorial videos prepared by the instructor to guide in the building of models were shared privately through video messaging platforms according to the needs of each student. This allowed for asynchronous teaching that leveraged the benefits of having continuous access to key course materials. WhatsApp, a tool that is particularly suited for establishing ongoing contact both during and outside of class hours because it allows the sharing of images, audios, and videos, helped establish an exchange of ideas with graphic annotations between the student and the instructor, between the instructor and the class, or between peers. These visuals were accompanied with voiceover feedback for further elucidation of the sketches to accommodate students' particular needs.

The blended format also bridged the desk crit to studio culture, where the desk crit was transformed from individual feedback to one between the student and peers that is monitored by the instructor. Online desk crits encouraged other students to take part in discussions between instructors and students on projects that have significantly different research and theoretical bases. This kind of interaction empowered both the student presenting and those listening by heightening the collaborative spirit for the success of each individual project. This emerging practice, where competitiveness of desk crits became collaboration between students [2], allowed for forging connections between students and helped in improving the quality of individual student outputs.

For the final review presentations, we have observed that students submit eloquent audio presentations when they must present components of their work online, but fail to do so when creating the same presentation for an in-person review panel. This likely happens because students tend to manage their time differently when in-person.

4 Conclusions

Students adapt well in an online environment to the theoretical component of the studio, to conducting research on the studio brief that outlines the overarching theme of the term project, and to preparing digital presentations for juries. These elements can be explored in a direct or flipped environment. Students can also take advantage online of having more time for their own research and reflection as they work on transcribing their research into sketches and models and receive regular feedback via remote sessions or the use of social media.

To a certain extent, skill development can take place online through video conferencing or social media feedback, which can include interaction with peers and instructors. For aspects that relate to the needed tactile skills for both critical thinking and visual creativity, however, there is a need for in intensive in-person sessions strategically distributed throughout the term, both for the development of these skills and for the execution of the final drawings and models.

References

1. Bachelard, G.: The Poetics of Space, 1994 edn. Beacon Press, Boston (1994)
2. Bender, D., Vredevoogd, J.D.: Using online education technologies to support studio instruction. Educ. Technol. Soc. **9**(4), 114–122 (2006)
3. Crowther, P.: Understanding the signature pedagogy of the design studio and the opportunities for its technological enhancement. J. Learn. Des. **6**(3), 18–28 (2013)
4. Fleischmann, K.: Hands-on versus virtual: reshaping the design classroom with blended learning. Arts Human. High. Educ. **20**(1), 87–112 (2021)
5. Haddad, E.G.: Christian Norberg Schulz and the project of phenomenology in architecture. Archit. Theory Rev. **15**(1), 88–101 (2010)
6. Hokanson, B.: The design critique as a model for distributed learning. In: Moller, L., Huett, J.B. (eds.) The Next Generation of Distance Education, pp. 71–83. Springer, Boston (2012). https://doi.org/10.1007/978-1-4614-1785-9_5
7. Lee, N.: Design as a learning cycle: a conversational experience. Eval. Innov. Dev. **3**(2), 12–22 (2006)
8. Merleau-Ponty, M.: The Visible and the Invisible. Northwestern University Press, Evanston (1968)
9. Mohammed, M.: Blended e-learning in the architectural design studio: an experimental model. Int. J. Parallel Emergent Distrib. Syst. **32**(suppl.1), 73–81 (2017)
10. Tessier, V., Aubry-Boyer, M.-P.: Turbulence in crit assessment: from the design workshop to online learning. Design Technol. Educ. Int. J. **24**(4), 86–95 (2021). Jones, D., Lotz, N. (eds.)
11. Tüfek, T.E.: An unexpected shift to an online design studio course: student insights on design critiques. Int. J. Art Design Educ. **41**(1), 158–170 (2022)

Is It Possible to Improve the Development of Executive Functions in Children by Teaching Computational Thinking?

Carolina Robledo-Castro[1,2](✉), Luis Fernando Castillo-Ossa[2,3,5], and Christian Hederich-Martínez[2,4]

[1] Universidad del Tolima, CUS Research Group, Calle 42 1-02, 730006 Ibagué, Colombia
crobledoc@ut.edu.co

[2] Universidad Autónoma de Manizales, Artificial Intelligence Research Group, Antigua Estación del ferrocarril, 170002 Manizales, Colombia

[3] Universidad de Caldas, Calle 65 26-10, 170002 Manizales, Colombia
luis.castillo@ucaldas.edu.co

[4] Universidad Pedagógica Nacional, Calle 72 11-86, Bogotá, Colombia
hederich@pedagogica.edu.co

[5] Universidad Nacional de Colombia Sede Manizales, Campus la Nubia, Manizales, Colombia

Abstract. Multiple studies have investigated the impact of teaching coding on some cognitive processes, with promising results. Based on this background, the current study evaluated the effect of a computational thinking intervention on the executive functioning of school-age children. The research had a between-subjects experimental design, with pre and post measurements and a control group. The measurement instrument was the BANFE-2 battery. The experimental group participated in an 8-week intervention in which they were taught to coding using the micro:bit device and the MakeCode programming environment. The findings revealed significant transfer effects on the executive functions of the students.

Keywords: Computational thinking · Coding · Executive functions · Educational intervention

1 Introduction

Computational thinking (CT onwards) refers to the set of cognitive processes involved when solving a problem computationally [1]. The steps recognized in a standardized way in CT include abstraction, decomposition, modeling, algorithm design, execution and correction [1–3]. Programming and coding are process involved in CT. Learning CT refers to developing an approach to problem-solving whose ultimate goal is to provide a solution to be programmed on a computer [4].

There exists a growing interest in introducing the principles of CT in K-12 education [5, 6] as an opportunity to develop logical thinking, problem-solving, and creative designing processes. Other studies investigated the cognitive processes involved in CT,

M. Temperini et al. (Eds.): MIS4TEL 2022, LNNS 580, pp. 7–12, 2023.
https://doi.org/10.1007/978-3-031-20617-7_2

those found strong links between processes such as mathematical thinking, language processing, attention, among others [7, 8].

Another area of study has focused on investigating how programming learning promotes the development of higher mental functions [1, 9]. In this regard, some systematic reviews have collected related studies regarding the topic here mentioned and have examined the transfer effects evidenced in such studies [9]. The authors found that the learning of CT has essential effects on improving some cognitive processes, especially in mathematical thinking, critical thinking, creativity, metacognition and reasoning.

These results are consistent with other studies that have found important effects of teaching CT on academic performance, especially in mathematics, and some cognitive processes, such as visuospatial reasoning, verbal reasoning [10], numerical skills, sequencing skills [11], fluid intelligence and creative thinking [12].

A few recent studies have been interested in executive functions such as inhibition [13, 14], planning [14–16] and working memory [15, 16]. Executive functions (EFs onwards) refer to cognitive processes conducted by the prefrontal cortex of the brain, associated with the organization of conscious activity and metacognition; in other words, they are those that allow organizing, programming, regulating, making flexible, and verifying mental activity and behavior, especially goal-directed [17, 18].

In conclusion, evidence has been found on how learning coding and the development of CT contribute to strengthening cognitive and neurological development in children. However, despite the findings and the relevance of this topic, research on the subject is still scarce. In particular, few studies have evaluated the effect of teaching CT on executive functions. Through this study, we have proposed investigating the effect of a CT educational program on the EFs of school-age children. This work aims to be a preliminary input for a randomized clinical trial of greater impact.

2 Materials and Method

Thirty fifth-grade children age 10 to 11 of a primary school in Colombia participated in the study. They were randomly assigned to a group experimental (N = 17) or a control group (N = 13) by an external teacher to the research. This study had a between-subjects experimental design with pretest and posttest measurements and control group. The study complied with all the ethical aspects of the Helsinki declaration and was endorsed by the bioethics committee of Universidad Autónoma de Manizales (act 124, September 29, 2021). Also, informed consent was signed by the participants and their families who agreed to participate in the study.

2.1 Outcome Measures

For the pre and post measurements, the BANFE-2 neuropsychological battery of executive functions and frontal lobes was selected [19]. The battery has a variety of cognitive processes related to executive functioning and the prefrontal lobes: Stroop test, "Iowa" card test, Labyrinths, Hanoi tower, Wisconsin Card Sorting Test, Generation of semantic classifications, Comprehension and selection of proverbs, Metamemory curve, Generation of verbs, Self-directed pointing, Consecutive subtraction, Sequential visuospatial and Verbal working memory.

All the tests that make up the battery have a wide background in the scientific literature and are area-specific, determined by both clinical studies and functional neuroimaging studies. Therefore, its application guarantees the generalization and comparison of results between different populations. The tests have an agreement between applicators of .80, which shows a high-reliability index. Also empirical studies have widely supported the consistency and construct validity of each test. Although each test can be interpreted individually, the global performance index was implemented for this study.

2.2 Intervention

The intervention received by the experimental group consisted of an eight-week CT program, with two sessions per week, of approximately 120 min each, which were carried out in the school's computer rooms. The control group continued to receive their computer classes normally.

During the first five weeks, the sessions of the project "programming for children" of the Ministry of Information Technologies and Communications [20] were developed. For the last three weeks, the researchers designed projects in which the children had to solve problems computationally [21]. The devices used in the intervention were computers and a micro:bit microprocessor (a small programmable card designed by the BBC that integrates numerous sensors). Each session included plugged and unplugged activities; for these last ones, the children learned to program in block language through the MakeCode platform and later verified their execution in the micro:bit device (Table 1).

Table 1. Intervention program for weeks

Weeks	Learning and topics
1	Basic concepts on CT
2	Micro:bit caracteristics. Introduction to the use of MakeCode editor
3	To interprete fuid diagrams, to use conditionals and micro:bit sensors, and to program outputs through LEDs
4	To communicate instructions through LEDs, to use Boolean variables and logical operations to make decisions
5	To define internal variables and to perform operations with them
6–8	Work based on projects: to design games and materials: a compass, a clock, dice, a humidity reader, a pedometer, and a temperature controller device

3 Results

The descriptive statistics of the study show that both the control group and the experimental group had a similar mean age. Table 2 shows the means and the standard deviations of the score obtained.

Table 2. Descriptive statistics by group

	Total N = 30		Control N = 13		Experimental N = 17	
Age in months						
- Mean (SD)	131,2 (5,7)		130,9 (4,8)		131,3 (6,49)	
Gender						
- Female (%)	8 (26,6)		3 (23,1)		5 (29,4)	
- Male (%)	22 (77,3)		10 (76,9)		12 (70,6)	
Total score BANFE-2						
	T1	T2	T1	T2	T1	T2
- Mean	362,4	390,8	365.3	379.7	360.2	399.3
- SD	25,9	22,2	23.4	23.9	28.9	17.8

Note: SD: Standard deviation; T1: pretest score; T2: posttest score

We applied an Analysis of Variance of Repeated Measures (ANOVA RM). The intra-subject factor of repeated measures was the pretest and posttest measures; the itersubject factor was the difference between the experimental and control groups. Due to the sample size (N = 30), we selected the levels of ω^2 to assess the effect size. The verification of the assumptions of Levene test confirmed equality of variances, normality of the score, and homoscedasticity for application of the ANOVA RM test both in the pretest $F(1,28) = 2.18$ $p = .150$ and as in the posttest $F(1,28) = 1.86$ $p = .184$.

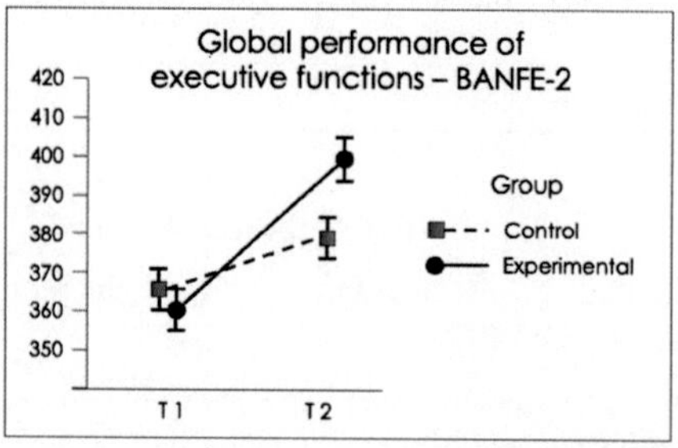

Fig. 1. Mean of the performances and standard errors of the groups.

The test means an increase between the pretest and the posttest for both groups, but this increase is greater in the experimental group (Fig. 1). Even though the overall performance of EFs started in the pretest of the experimental group with lower scores than the control group, this group showed higher performances in the posttest. Despite there being no global differences between the groups $F(1,28) = 0{,}77$ $p =,388$ $\omega^2 =,00$, there is a significant global difference between the pretest and the posttest with a very large effect size $F(1,28) = 85{,}15$ $p <,001$ $\omega^2 = 0{,}24$ and a significant difference, with medium effect size, linked to the interaction between the time of the test and the group $F(1,28) = 18{,}11$ $p <,001$ $\omega^2 = 0{,}06$. In conclusion, the intervention seems to have caused moderate to large changes in the total scores of the BANFE-2 battery.

4 Discussion

The pretest and posttest results showed that the experimental group significantly improved their performance in tasks related to EFs and the prefrontal cortex compared to a group of equivalent age and context that did not participate in the intervention. These differences were maintained when statistical corrections were made, showing a medium to large effect of the intervention on the global performance of executive functioning. These findings suggest that an 8-week CT intervention, developed with micro:bit microprocessors, may positively affect cognitive processes related to executive functions and prefrontal functioning.

These results are consistent with Robertson's research [16] who found a relationship between EFs and creative programming and debugging processes related to CT. The authors found that a CT intervention with the programming language in blocks with Scratch had a significant effect on the general development of EFs in children aged 10 to 12. That study used the Behavior Rating Inventory of Executive BRIEF-2, a neuropsychological test based on behavioral reports from adults. In our research, however, the BANFE-2 test battery was implemented, which is composed of classic neuropsychological tasks that are based on the execution of the subject. Despite the difference between the type of primary measures used, both studies agree on the relationship between both variables studied, computational thinking and EFs.

Such findings are also similar to the results of other studies that have evaluated the effect of interventions in CT on some specific EFs, evaluated with classical neuropsychological tasks such as inhibitory control [17, 18], working memory [18, 19] and planning [17, 19]. However, most of these studies have been conducted in the preschool population.

In relation to the intervention, one of the strengths of the program was the incorporation of connected and disconnected activities; this exercise favored the children's learning of concepts of a high level of abstraction; on the other hand, the use of the micro bit for projects such as a pedometer, a compass, and a humidity reader, allowed the students to recognize the usefulness of coding for solving problems in everyday life.

5 Conclusions

The intervention program aimed at teaching CT and programming in block language in school children aged 10 to 11 demonstrated to have a moderate to large effect on the global development of their executive functions and prefrontal cortex.

One of the study's limitations is the small sample (N = 30) and the ack of blinding of allocation and assessors. However, the results obtained in this pilot test show the relevance of replicating this study through research that complies with the protocols for a randomized clinical trial. For future studies, it is suggested to include a measure to assess children's CT in both the pretest and the posttest. It is also suggested to extend the duration of the intervention and incorporate the principles of cooperative learning in the design of the activities.

Funding. This research was financed by the Research Fund of the University of Tolima, project 430121.

References

1. Wing, J.: Computational thinking and thinking about computing. Philos. Trans. R. Soc. A: Math. Phys. Eng. Sci. **366**(1881), 3717–3725 (2008)
2. Shute, V.J., Sun, C., Asbell-Clarke, J.: Demystifying computational thinking. Educ. Res. Rev. **22**, 142–158 (2017)
3. Kong, S.-C., Abelson, H. (eds.): Computational Thinking Education. Springer, Singapore (2019). https://doi.org/10.1007/978-981-13-6528-7
4. Beecher, K.: Computational Thinking: A Beginner's Guide to Problem- Solving and Programming. The Chartered Institute for IT (2017)
5. Zhang, L., Nouri, J.: A systematic review of learning computational thinking through Scratch in K-9. Comput. Educ. **141**, 103607 (2019)
6. Sun, L., Hu, L., Zhou, D.: Improving 7th-graders' computational thinking skills through unplugged programming activities: a study on the influence of multiple factors. Thinking Skills Creativity **42**, 100926 (2021)
7. Floyd, B., Santander, T., Weimer, W.: Decoding the representation of code in the brain: an fMRI study of code review and expertise. In: IEEE/ACM 39th International Conference on Software Engineering, ICSE 2017, pp. 175–186 (2017)
8. Ivanova, A.A., et al.: Comprehension of computer code relies primarily on domain-general executive brain regions. Elife **9**, 1–24 (2020)
9. Scherer, R., Siddiq, F., Sánchez-Viveros, B.: The cognitive benefits of learning computer programming: a meta-analysis of transfer effects. J. Educ. Psychol. **111**(5), 764–792 (2019)
10. Strawhacker, A., Bers, M.U.: What they learn when they learn coding: investigating cognitive domains and computer programming knowledge in young children. Educ. Tech. Res. Dev. **67**(3), 541–575 (2019). https://doi.org/10.1007/s11423-018-9622-x
11. Tsarava, K., et al.: A cognitive definition of computational thinking in primary education. Comput. Educ. **179**, 104425 (2022)
12. Wang, L., Geng, F., Hao, X., et al.: Measuring coding ability in young children: relations to computational thinking, creative thinking, and working memory. Curr. Psychol. (2021) https://doi.org/10.1007/s12144-021-02085-9
13. Robertson, J., Gray, S., Martin, T., Booth, J.: The relationship between executive functions and computational thinking. Comput. Sci. Educ. Schools **3**(4), 35–49 (2020)
14. Arfé, B., Vardanega, T., Rononia, L.: The effects of coding on children's planning and inhibition skills. Comput. Educ. **148**, 2–16 (2020)
15. Di Lieto, M.C., et al.: Empowering executive functions in 5- and 6-year-old typically developing children through educational robotics: an RCT study. Front. Psychol. **10**(1), 3084 (2020)
16. Gerosa, E., Koleszar, V., Gómez, L., Tejera, G., Carboni, A.: Cognitive abilities and computational thinking at age 5: evidence for associations to sequencing and symbolic number comparison. Comput. Educ. Open **2**, 100043 (2021)
17. Diamond, A.: Activities and programs that improve children's executive functions. Curr. Dir. Psychol. Sci. **21**(5), 335–341 (2013)
18. Goldstein, S., Naglieri, J.A.: Handbook of Executive Functioning. Springer, Cham (2013)
19. Flores-Lázaro, J.C., Otrosky-Shejet, F., Lozano-Gutierrez, A.: BANFE 2 - Batería neuropsicológica de funciones ejecutivas y lóbulos frontales. Manual Moderno (2014b)
20. Ministry of Information Technologies and Communications 2021. Programación para niños y niñas (2021). https://talentodigital.mintic.gov.co/734/w3-article-162055.html
21. Kay, J., et al.: Problem-based learning for foundation computer science courses. Comput. Sci. Educ. **10**(2), 1–20 (2000)

Supporting the Semi-automatic Feedback Provisioning on Programming Assignments

Sven Strickroth(✉) and Florian Holzinger

Institute for Informatics, Ludwig-Maximilians-Universität München, Oettingenstraße 67, 80538 München, Germany
sven.strickroth@ifi.lmu.de

Abstract. Feedback is important for learning, however, manual feedback provisioning is time and resource consuming. For programming education, various systems have been developed to automate recurring tasks or the whole feedback generation. As fully automated systems are not always a viable option, this paper investigates how teaching assistants can be supported in semi-automated assessment of programming assignments to give good feedback more easily. An existing semi-automated e-assessment system is extended with configurable feedback snippets as well as adaptively recommended feedback snippets based on feedback of similar submission in order to evaluate the effects on the grading and the feedback given by teaching assistants. The results indicate that such feedback snippets lead to more consistent and motivational feedback, can help finding mistakes, and have no impact on the awarded grades.

Keywords: Feedback provisioning · Assessment · Usability

1 Introduction

A central component of introductory programming courses in higher education are homework assignments in which students write small programs themselves or extend existing programs on a regular basis. As learning programming is often perceived as hard by novices, a timely, individual, explanatory, and motivational feedback is important [3,8,9]. Fully automated systems can provide instant and objective feedback [6], but they are still limited as all tests and feedback need to be pre-defined and not all aspects such as creativity or implementation quality can be assessed automatically [7]. Therefore, manual or semi-automated grading is often necessary. This requires a significant amount of time and effort that only scales with the number of teaching assistants. There is, however, usually a (monetary) limitation on how many assistants can be hired and having more teaching assistants is likely to introduce inhomogeneities in feedback and grading.

This paper investigates the research question of how to support teaching assistants in semi-automatic assessment so that they can provide consistent and

M. Temperini et al. (Eds.): MIS4TEL 2022, LNNS 580, pp. 13–19, 2023.
https://doi.org/10.1007/978-3-031-20617-7_3

motivating feedback more easily. The main idea is to automate recurring tasks and to provide pre-defined feedback snippets as well as adaptively recommended feedback snippets based on similar submissions that have already been assessed.

The remainder is organized as follows: First, the state of the art is presented. This is followed by descriptions of the prototype, the evaluation design, and the results. The paper closes with a discussion, a summary, and an outlook.

2 Exploring the Design Space: State of the Art

There is a large number of systems and approaches to support teachers in programming education in terms of manual and semi-automatic grading and providing feedback [6,11]. All supportive systems collect the students' solutions at a central place and allow teaching assistants to download or view these.

There seem to be two (orthogonal) dimensions on which systems focus: First, there are semi-automated systems such as Praktomat [2] that automate recurring tasks, such as compiling and testing of submissions, and present the results for manual grading to the teaching assistants. Such automation has shown to be very helpful, e.g. to identify mistakes easily, and a (significant) time saver [10]. Second, there are systems such as Rubyric [1] that focus on supporting the manual grading and feedback provisioning by providing a (detailed) form/rubric. The scoring rubrics can be either static or configurable and some tools (e.g., Rubyric) allow teachers to define re-usable feedback snippets. Rubyric was found helpful by teaching assistants, feedback was more consistent, and students were satisfied with amount and quality of the feedback [1].

Based on the experience that novice programmers often make similar mistakes or have common misconceptions, there is a concept proposal to further support the feedback provisioning by an collaborative approach [4]: During the assessment of a submission, a system could adaptively suggest already entered feedback for similar submissions that can be reused and personalized in order to release teaching assistants from writing the same feedback repeatedly.

To summarize: In all systems known to the authors there is no system that supports both dimensions and also provides adaptive recommendations for feedback snippets. Also, the authors are not aware of any evaluation of the effects of such supportive features on the awarded grades and the provided feedback.

3 Adaptive Feedback Snippet Recommendations

The research gap identified in the previous section is addressed by an extension of the e-assessment system GATE, which is in use since 2011 [10]. GATE supports teaching assistants by displaying the results of tests that are automatically executed after the submission deadline, a teacher configurable rubric, and (optionally) with plagiarism detection by linking similar submissions.

For this research GATE was extended with teacher configurable feedback snippets and an adaptive feedback snippet recommendation that is based on a simple clustering approach [5]: The outputs of all failed automated tests are parsed to detect Java exceptions, failed unit test steps, or syntax errors. Then, the similarities to all other failed test outputs are calculated and stored in the database. Finally, during the assessment, already given feedback for similar errors and test results is presented to the teaching assistants ordered by the calculated similarity. For technical reasons, the single-page assessment workflow needed to be split into a multi-page workflow so that the feedback snippets can be calculated for a specific assessment step (e.g. a syntax or unit test).

Figure 1 depicts an excerpt from the GUI for the teaching assistants of the new interface. The assessment consists of three steps/pages (an overview of the steps can be seen at the upper right: for the first two steps automatic tests are linked and only the first one passed) and a separate overview page. In the middle there is the form for awarding grades and leaving feedback for the current step. In the figure the current step concerns the syntactically correctness and also shows the result of the automatic syntax test (PASSED, otherwise the error would be shown). Grades are automatically pre-filled when all tests of a step were passed. Additionally, misconceptions can be created or added to a submission (cf. [4,5]). Directly below the form it is possible to select and insert a feedback snippet from a set of pre-defined or adaptively recommended feedback snippets. Always displayed at the bottom are the files submitted by the student (with syntax highlighting).

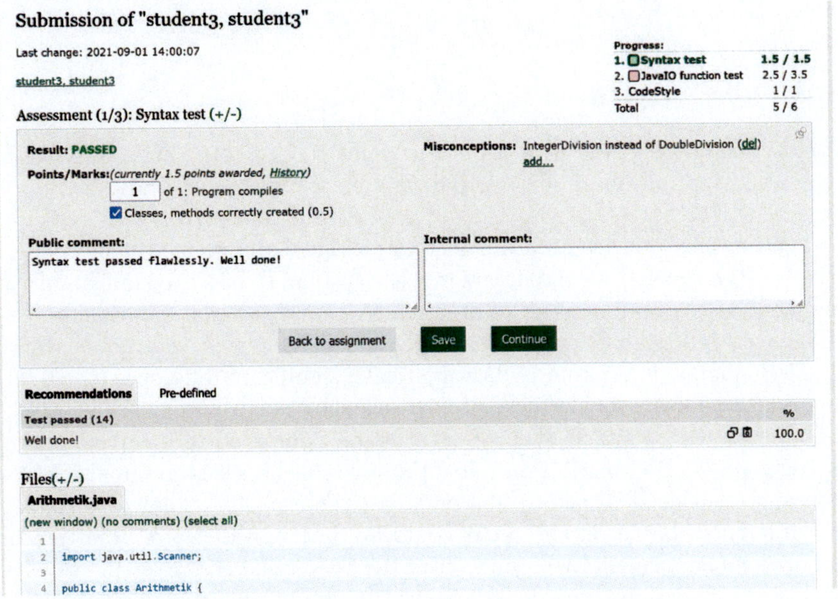

Fig. 1. User interface of the assessment interface of the prototype consisting of three assessment steps/pages, showing the first step *(translated to English)*.

4 Evaluation

To investigate the effects on the grading and given feedback as well as the usability, an empirical lab study with a within-subjects design and a think-aloud approach was conducted. Fifteen participants took part in the study (f = 5, m = 10) with an age between 23 and 55 (mean = ø = 28.3, median = 26). The participants were randomly split into two groups which either used the "old" system ($A \rightarrow B$) or the new prototype ($B \rightarrow A$) first to address ordering effects. All participants were either students (10) or teachers (5) from three German universities. Eleven participants already had experience as teaching assistants whereas four of these rated their teaching experience as little to medium and seven as much to very much. Nine participants were part of the teaching team for a bachelor's course on introductory programming at least once and three participants were part of the teaching team of programming practicals at least once. Two of the participants had experience with the GATE system and five have used other assessment tools.

For the study two mock courses each consisting of 20 students, the same programming assignment, and (comparable) submissions were constructed (based on real ones). For each course four submissions were not assessed, yet, of which two contained a mistake. During the study the participants got a quick introduction into the prototypes and then had the task to assess these eight submissions. The time needed for completing the assessment was measured. At the end a semi-structured interview was conducted. The study used Zoom with screen sharing and was recorded. Two 50 Amazon vouchers were raffled among the subjects.

5 Results

All participants were able to use both systems to assess the eight submissions and it took approx. one hour per participant to finish the study. Except for one person (P12, $A \rightarrow B$) who did not spot a functional error with system A, all other participants detected all mistakes in the submissions to be assessed. However, P12 spotted a similar error with system B based on an adaptively recommended feedback snippet of an already assessed similar submission.

The average length of the feedback given with system A was 103 characters and 181 characters for system B – a statistically significant difference ($p = 0.001$). An in-depth analysis for correct submissions shows that the given feedback is much shorter (A: ø = 52, B: ø = 158 characters) compared to incorrect submission (A: ø = 133, B: ø = 195) and that the given feedback is minimal longer in the first system used (ø = 160) compared to the second system (ø = 125). No other differences based on the order could be determined. Note, however, that there are two participants who once gave the very same comment in two steps. These two duplications have been resolved before calculating the quantitative statistics above. The qualitative analysis of the given feedback shows that in all comments the mistakes were clearly named except for the one participant (P12) who did not spot an error with system A. Independent of the system used for giving the feedback but dependent on the personal assessment style, some participants gave hints on how to resolve the mistake. Overall, the feedback given using system B seems to be more consistent/homogeneous and contains more positive/motivational comments compared to feedback given using system A (A: 23, B: 48). The awarded grades show no (statistically significant) difference between the two systems (A: $\varnothing = 4.7$; B: $\varnothing = 4.9$, $p = 0.32$; grades from 0 worst to 6 best).

The time required for the assessment of the submissions shows a statistically significant difference ($p = 0.048 < 0.05$): The assessment with system A took 3:29 min. on average and with system B 4:13 min. (a difference of 21 %). Also, the assessment of correct submissions took about one minute (i. e. 33 %) longer with system B compared to system A. When system A was used first, the correction of incorrect submissions took about the same time ($\varnothing = 4$ min.) for both systems. When system B was used first, however, the assessment of incorrect submissions was about 44 % quicker with system A (A: $\varnothing = 3{:}27$, B: $\varnothing = 4{:}59$ min.). Correct submissions were always assessed quicker than incorrect ones.

In the interviews 80% of the participants explicitly stated a preference of system B. System A was rated three times as not as supportive as system B. The feedback snippets were mentioned positively 7 times and were estimated to result into more consistent feedback by two participants. Three participants each stated that they felt they provided more positive feedback or more feedback overall with system B. However, two participants noted that the additional features also increase the overhead and lead to a longer time to get familiarized with system B. Five participants liked that the grades were pre-filled for steps in which all tests were passed and four participants rated the recommended feedback snippets as particularly helpful for spotting the cause of the mistakes. Additionally, three students missed a feature to directly correct a syntax error and being able to re-run the (unit) tests within the system.

Regarding the workflow, 12 participants rated the multi-step version as good but system A was classified as more efficient compared to system B by six participants. One participant (P6) explained this by saying everything is visible at a glance and this is particularly helpful as "assessment means multitasking". Three participants noted however that they were less afraid of making mistakes in system B, the feedback quality increases, the steps help structure the feedback, and allow to focus on one aspect at a time. The detailed usability analysis and usage of the misconception feature are skipped here due to page restrictions.

6 Discussion, Conclusions and Outlook

This paper investigated how teaching assistants can be supported in semi-automatic assessment to give feedback more easily. A prototype was built that automates recurring tasks and provides pre-defined feedback snippets as well as adaptively recommended feedback snippets based on similar submissions. The results of an empirical lab study indicate that the feedback snippets have been very appreciated by the teaching assistants and lead to more consistent and motivational feedback without influencing the awarded grades. The snippets also seem to be particularly helpful for detecting causes of mistakes and do not result in many duplicate comments. The time needed for an assessment, however, seems to be higher for the prototype. This was particularly true for the first submissions – the last ones were assessed much quicker (about 40 %; 12 % for the old system). Reasons could be that more clicks are needed for a single assessment but it is also possible that it is caused by the longer feedback or that the distinct steps triggered a deeper investigation of the submissions. Interestingly, the objective time measurement contradicts the subjective perception of five participants of being quicker with the prototype. Note, however, that the time measurement might be biased due to the think-aloud approach but should still show a tendency.

The main limitation of this research is the lab study with eight assignments to be assessed. This number might seem small, however, the time (one hour) per participant for the study was the limiting factor. Still, the participants came from different universities and cover a wide range of different experiences.

Pre-defined and adaptively recommended feedback snippets are likely to be transferable to (both formative and summative) assessments in STEM to improve the given feedback, even if automatic tests are not or only partially available, and might also be helpful in settings such as peer review and peer feedback.

The workflow needed to be changed from a single-page to a multi-page assessment for the adaptive feedback snippet recommendation. Several participants stated that they preferred either of the two workflows. Due to the study design no clear conclusions can be drawn upon the multi-page vs. single-page workflow. Therefore, further research is needed to compare both approaches with the same feature set and the procedures of how solutions are assessed.

References

1. Auvinen, T.: Rubyric. In: Proceedings of Koli Calling. ACM Press (2011). https://doi.org/10.1145/2094131.2094152
2. Breitner, J., Hecker, M., Snelting, G.: Der Grader Praktomat. In: Automatisierte Bewertung in der Programmierausbildung, pp. 159–172. Waxmann (2017)
3. Butler, M., Morgan, M.: Learning challenges faced by novice programming students studying high level and low feedback concepts. In: Proceedings Ascilite Singapore, pp. 99–107 (2007)
4. Heller, N., Bry, F.: Human computation for learning and teaching or collaborative tracking of learners' misconceptions. In: Intelligent Systems and Learning Data Analytics in Online Education, pp. 323–343. Academic Press (2021). https://doi.org/10.1016/B978-0-12-823410-5.00015-2
5. Holzinger, F.: Kollaborative Unterstützung bei der semi-automatischen Bewertung von Programmieraufgaben. Master's thesis, LMU Munich, Germany (2021)
6. Keuning, H., Jeuring, J., Heeren, B.: A systematic literature review of automated feedback generation for programming exercises. TOCE **19**(1), 1–43 (2018). https://doi.org/10.1145/3231711
7. Laß, C., et al.: Stager: simplifying the manual assessment of programming exercises. In: Proceedings of SEUH, vol. 2358, pp. 34–43. CEUR-WS (2019)

8. Luxton-Reilly, A., et al.: Introductory programming: a systematic literature review. In: Proceedings of ITiCSE, pp. 55–106 (2018). https://doi.org/10.1145/3293881.3295779
9. Moreno, R.: Decreasing cognitive load for novice students: effects of explanatory versus corrective feedback in discovery-based multimedia. Instr. Sci. **32**(1), 99–113 (2004). https://doi.org/10.1023/B:TRUC.0000021811.66966.1d
10. Strickroth, S., Olivier, H., Pinkwart, N.: Das GATE-System: Qualitätssteigerung durch Selbsttests für Studenten bei der Onlineabgabe von Übungsaufgaben? In: Proceedings of DeLFI, pp. 115–126. Bonn, Germany (2011)
11. Strickroth, S., Striewe, M.: Building a corpus of task-based grading and feedback systems for learning and teaching programming. iJEP **12**(5) (2022). https://doi.org/10.3991/ijep.v12i5.31283

Educational Chatbot to Support Question Answering on Slack

Simone Leonardi(✉) and Marco Torchiano

Politecnico di Torino, Corso Duca degli Abruzzi 24, 10129 Turin, Italy
{simone.leonardi,marco.torchiano}@polito.it

Abstract. Educational chatbots unlock new possibilities to support teaching activities. In recent years, the progress made in natural language processing and understanding allowed the development of virtual assistants that understand questions written as unstructured data. They generate a proper answer based on knowledge bases. This work proposes an accurate solution easy to test in real world scenarios like university courses. In fact, the use of Slack and RASA technologies reduces the difficulties to train and install the educational chatbot on already existing digital workplaces. We evaluates the effectiveness of our solution reaching 0.93 accuracy in intent recognition building a stacked neural network.

Keywords: Educational chatbot · Question answering · NLP · Remote teaching

1 Introduction

The COVID-19 pandemic increased the needs for digital tools also in the educational system. Students attended online classrooms with remote teachers. In this context universities look for effective digital supports to continue offering their services without losing the interaction with the students. It has been shown that the adoption of Artificial Intelligence in education is rapidly expanding [7]. In detail, one of the most popular AI tool in education is the chatbot system [4]. A chatbot is a virtual assistant able to interact with human inputs, process it and produce an answer and it also promotes further development in the conversation. In a scenario where students need to interface with professors remotely, there is the possibility to support teachers in answering questions sometimes recurrent or already answered in the previous edition of the course. An educational chatbot is the ideal support for this specific task. In [2], Cunningham et al. propose practical steps to prototype and to develop such tools. In the recent years several studies have demonstrated how the adoption of chatbot in education leads to positive outputs in the personalization of contents offered and in the learning process [1,6,8]. In this work we focus on the ease of deployment of a solution to be installed on Slack, a widespread digital workplace that allows interaction between users. We select RASA Open Source as a server side environment because it is easy to configure its NLU (Natural Language Understanding)

M. Temperini et al. (Eds.): MIS4TEL 2022, LNNS 580, pp. 20–25, 2023.
https://doi.org/10.1007/978-3-031-20617-7_4

architecture and it already includes tools to enlarge the Q&A dataset directly in the Slack channel when submitting a new answer. We developed this work by answering these research questions: *RQ1*: Which dataset can be used to train and test a chatbot in the educational system? *RQ2*: Which natural language understanding model is the most accurate in question answering for the educational context? *RQ3*: Is a chatbot deployed on a common chat platform able to understand a student's intent during a multi turn conversation? The paper is structured as follows: in Sect. 2 we review related work, while in Sect. 3 we describe the proposed approach for training and classifying students' questions. In Sect. 4 we present a preliminary experimental evaluation of our method. Eventually, in Sect. 5, we conclude with a summary of the experiment outcome and outlining future work.

2 Related Work

The role of chatbot in education have been widely explored in the recent years leading to a large number of publications [1,2,4–8]. Students interact with chatbot systems to receive immediate answers to common questions even when the professor is not available or it is difficult to retrieve the information on the course material. In this work we offer a solution for the students that lack the assistance of professors in large courses, especially those that are held online, providing a conversational agent able to immediately answer a question based on previous answers given by professors. On the other side this agent is able to learn new answers by interacting directly with professors. We compare our solution with different baseline models to assess its accuracy. The final pipeline is described in Sect. 3, where we define a client-server infrastructure to solve this issue. The model is expandable with Q&A of the same topic or by changing the training dataset with different school subjects. Our educational chatbot also allows a fallback procedure to forward the question to a human assistant.

3 Approach

In the following we explain the data retrieval phase, the model implementation, and the interaction pipeline.

3.1 Dataset

The questions and the answers collected came from the Object Oriented Programming course held during the COVID-19 pandemic at Politecnico di Torino in the spring semester 2020. All the students' information have been anonymized for privacy reasons. This course provided remote assistance with the support of the Slack web application. The workspace of the course has been divided into two main channels: theory and laboratories. The first channel contains questions about frontal theory lessons while the second channel is about the exercises given during the practical sessions of the course. The topic selected for the experiment is the *inheritance* with its 53 subtopics and 300 questions answered. This is the training set of our model while for the testing part, we used the same topic questions from the 2021 edition of the same course. The intent plays the role of labels in the context of supervised classification. All the questions have been cleaned by removing emojis, url and references to previous questions in the channel with tags and, similarly, all the handles to cite a user in the channel have been deleted. Finally questions have been labeled as *intents* while answers have been labeled as *actions*. Questions and answers are paired following a specific format[1]. This dataset answers our first research question RQ1 and it also explains how to build a dataset to be processed in our model.

3.2 Question Answering Model

The neural network model adopted in our solution is ConveRT (Conversational Representations from Transformers). Henderson et al. in [3] propose this Transformer based dual encoder specialized in conversational tasks. In Fig. 1, we illustrate the single-context ConveRT model that is the building block for the multi-context solution. In fact, only the multi-context solution is able to handle multi turn conversation. In our use case scenario, students can interact with the chatbot asking for further clarification after receiving the first answer. Similarly, they can ask multiple questions before closing the conversation. We decide to include the multi-context ConveRT in our solution because it is the best performing one with respect to the BERT and SpaCy[2] ones, as we demonstrate in the validation Sect. 4 and because it is faster on our cloud based solution.

The goal of ConveRT is to select the right answer once given the question. There are two alternatives: the question exists in our dataset or it does not exist. In any case, our model tries to match the question based on similarity and automatically print the paired answer. If the answer is not satisfactory for the student, he can rephrase the question or he can choose to forward the question to the professor.

3.3 Client Server Architecture

In this section we describe the architectural choices made to develop our educational chatbot and how to set up the environment to deploy it. We use Slack[3] application as a communication channel between students and professors. It plays the role of the web client. It also hosts our educational chatbot. In parallel, we use RASA Open Source on the server side of the communication to handle the natural language understanding processes.

We start the development of the bot creating a Slack application on https://api.slack.com/apps. We create a RASA project and we define how our context-

1 https://rasa.com/docs/rasa/stories/.
2 https://spacy.io/models.
3 https://slack.com/.

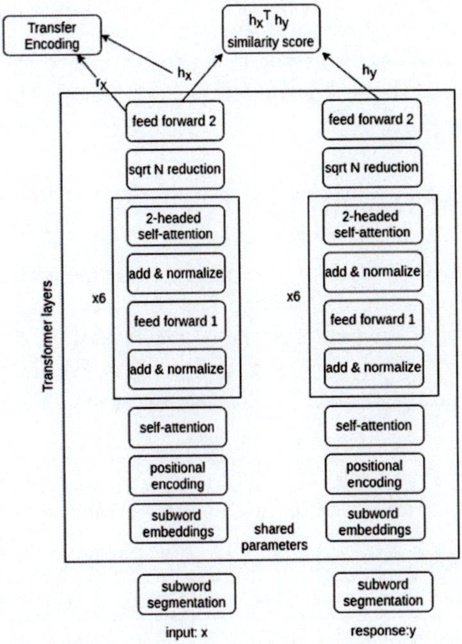

Fig. 1. Block diagram describing the actions performed from the initial question asked by the student on the Slack course channel through the RASA server intent recognition and search for an answer. This phase can be successful and return the answer on Slack by the chatbot or forward the question to the professor on Slack. Once finally answered the interaction is stored in the update dataset for future uses.

aware conversational chatbot will handle conversations. There are three main tasks to be configured: slots, stories and policies.

- *Slots*[4] are the assistant's memory. They store pieces of information that the bot needs to refer to later and can direct the conversation flow based on slot events.
- *Stories* are examples of conversations between a user and the bot. In stories different patterns of interaction are described.
- *Machine Learning Policies* help the chatbot to better predict the response in unseen conversation paths.

The final component of RASA is the RASA X component. It is a CDD (Conversion Driven Development) tool. It saves automatically all conversations and it uses them to improve the chatbot and its embedded NLU model.

The entire pipeline of usage of the system is the following. A student writes a question in the Slack channel where the educational chatbot is installed. The question is forwarded to the RASA server and it is processed to recognise the

[4] https://rasa.com/docs/rasa/domain#slot-types.

intent and consequently find the best answer. If the search is a failure, then the question is forwarded to the professor that, answering to the question, updated the dataset and increased the knowledge base of the chatbot. In any case, the student will receive the answer.

4 Experimental Results

We assess the effectiveness and precision of our framework computing metrics over the intent recognition functionality. The problem evaluated is a multi label supervised classification. We compare the recall, precision, f1-score and accuracy of three main models: BERT, SpaCy and ConveRT. We use 80% of the dataset as a training set, while the remaining 20% as a test set. The results are listed in Table 1.

Table 1. Evaluation metrics computed with a 80/20 train/test set setup for the task of intent recognition.

Model	Recall	Precision	f1-score	Accuracy
ConveRT	0.93	0.91	0.92	0.93
BERT	0.92	0.89	0.90	0.92
SpaCy	0.92	0.88	0.89	0.92

We also tested 174 different conversations containing 921 different actions with a 5 fold cross validation approach. RASA offers in its suite the possibility to simulate user inputs that test various combinations of actions and turns in conversations.

Obtained results answer RQ2 and RQ3 addressing both the task of intent recognition and the correct conversation decisional path choice. We show that the best performing model is ConveRT in both tasks that reach accuracy above 0.93 and f1-score above 0.92. These results are the one obtained with the Italian language because our dataset was in Italian, it is possible that using other languages BERT or SpaCy could beat the ConveRT model, in this case the switch between different configurations is already arranged. This intuition is tested by Wu and Dredze in [9]. They prove that pre-trained language models perform better when they adopt a language with lots of digital resources publicly available and correctly formatted. As an example, English has over twenty gigabytes of data while Italian just two and the Yoruba language less than 0.01.

5 Conclusion

This work presents a new chatbot system to support question answering in education, especially useful for courses with a large number of students and in remote teaching due to pandemics. We present an educational chatbot that is deployable on Slack for ease of use. The server side of our framework, developed in RASA Open Source and Rasa X proved its accuracy of 0.99 with the ConveRT Natural Language Understanding model outperforming the other baselines models both in intent recognition and in correct conversational decisions. We trained the model with the course Slack channel in 2020 and we tested it on questions made during 2021 reaching high accuracy in response identification. Further improvements of this work involve the adoption of the system on other topics and teaching courses, plus making a survey once deployed on real time interaction during an academic semester. *We finally thank Fiorentino Cairone for his contribution in the implementation of the educational chatbot.*

References

1. Clarizia, F., Colace, F., Lombardi, M., Pascale, F., Santaniello, D.: Chatbot: an education support system for student. In: Castiglione, A., Pop, F., Ficco, M., Palmieri, F. (eds.) CSS 2018. LNCS, vol. 11161, pp. 291–302. Springer, Cham (2018). https://doi.org/10.1007/978-3-030-01689-0_23
2. Cunningham-Nelson, S., Boles, W., Trouton, L., Margerison, E.: A review of chatbots in education: practical steps forward. In: 30th Annual Conference for the Australasian Association for Engineering Education (AAEE 2019): Educators Becoming Agents of Change: Innovate, Integrate, Motivate, pp. 299–306. Engineers Australia (2019)
3. Henderson, M., Casanueva, I., Mrkšić, N., Su, P.H., Wen, T.H., Vulić, I.: ConveRT: efficient and accurate conversational representations from transformers. In: Findings of the Association for Computational Linguistics: EMNLP 2020, pp. 2161–2174. Association for Computational Linguistics, November 2020. https://doi.org/10.18653/v1/2020.findings-emnlp.196. https://aclanthology.org/2020.findings-emnlp.196
4. Okonkwo, C.W., Ade-Ibijola, A.: Python-bot: a chatbot for teaching Python programming. Eng. Lett. **29**(1), 25–34 (2020)
5. Okonkwo, C.W., Ade-Ibijola, A.: Chatbots applications in education: a systematic review. Comput. Educ. Artif. Intell. **2**, 100033 (2021). https://doi.org/10.1016/j.caeai.2021.100033. https://www.sciencedirect.com/science/article/pii/S2666920X21000278
6. Ranoliya, B.R., Raghuwanshi, N., Singh, S.: Chatbot for university related FAQs. In: 2017 International Conference on Advances in Computing, Communications and Informatics (ICACCI), pp. 1525–1530. IEEE (2017)
7. Roos, S.: Chatbots in education: a passing trend or a valuable pedagogical tool? (2018)
8. Sinha, S., Basak, S., Dey, Y., Mondal, A.: An educational chatbot for answering queries. In: Mandal, J.K., Bhattacharya, D. (eds.) Emerging Technology in Modelling and Graphics. AISC, vol. 937, pp. 55–60. Springer, Singapore (2020). https://doi.org/10.1007/978-981-13-7403-6_7
9. Wu, S., Dredze, M.: Are all languages created equal in multilingual BERT? (2020). https://doi.org/10.48550/ARXIV.2005.09093. arXiv:2005.09093

A Case Study on Students' Opinions About Adaptive and Classical Tests

Anna Maria Angelone and Pierpaolo Vittorini(✉)

University of L'Aquila, 67100 L'Aquila, Italy
{annamaria.angelone,pierpaolo.vittorini}@univaq.it

Abstract. Assessment in education is an important task. Albeit being relatively old techniques, the adoption of adaptive (CAT) instead of classical testing (FIT) still poses several questions. According to the literature, a CAT test should be faster to complete, more precise in estimating the student's ability, and more engaging to take than FIT. On the other hand, a FIT test should be easier to prepare and complete, and less prone to anxiety than CAT. In this context, the paper initially reports on a study conducted with students (that used CAT and FIT) about different aspects of their assessments' experiences. The results show that students seem to prefer CAT, considered fairer, more practical and faster than FIT, independently of the different attitudes towards technology. Furthermore, students did not report more anxiety or more workload with CAT. Moreover, we also evaluate different methods to convert the ability measured with CAT into a grade, so to verify if the adopted method could have influenced the results of the questionnaire.

Keywords: Assessment · Computerised adaptive testing · Fixed-item testing

1 Introduction

Online assessment is the use of information technology to support various forms of assessment, in different contexts (e.g., educational, health, psychiatric and psychological) [13]. Within any online assessment system, question types may vary. For example, questions may include open-ended questions, true or false type questions, or multiple-choice questions [19].

Questions accepting either dichotomous (i.e., right/wrong) or multiple-choice answers are usually called items. The properties of tests made up of items have been largely studied in psychometry, under the Classical Test Theory (CTT) [11] or Item Response Theory (IRT) [15]. A test created according to the CTT assigns to an individual an observed test score given as the unweighted sum of responses to test items, plus a casual error component. Basically, the candidate's level of preparation is therefore expressed as the sum of the correct answers. Tests administered using CTT are commonly called Fixed-Item Tests (FIT). On the other hand, the IRT states that the individual's probability of giving a correct

M. Temperini et al. (Eds.): MIS4TEL 2022, LNNS 580, pp. 26–36, 2023.
https://doi.org/10.1007/978-3-031-20617-7_5

answer depends on the interaction between the individual's ability and several item parameters (e.g., difficulty, guessing). Based on IRT, Computerised Adaptive Testing (CAT) is instead a type of computer-based test that adapts the questions to the examinee's ability.

According to the literature, a CAT test should be faster to complete [2], more precise in estimating the student's ability [30,33], and more engaging to take than FIT [21]. On the other hand, a FIT test should be easier to prepare and complete, and less prone to anxiety than CAT [8,17]. Even if it has been widely documented that IRT/CAT are more accurate in assessing abilities than CTT/FIT, it is only used to a minimum extent in education [24].

In this paper, we focus on the opinion of students about using CAT or FIT. In particular, we investigated their preferences, the perceived fairness and efficiency, the required preparation, the different way in which questions are asked, and their anxiety during the assessment. Furthermore, we also present different methods to convert the ability measured with CAT into a grade, and whether the method adopted in our case study may represent a bias in the results. Accordingly, the paper tackles the following research questions:

- *RQ1.* Which are the students' opinions about FIT and CAT, in terms of fairness, efficiency, preparation, modality of administration, and anxiety?
- *RQ2.* What is the impact of the different methods of converting the ability measured with CAT into a grade, with respect to FIT?

The paper is organised as follows. Section 2 introduces background information on CTT/FIT, IRT/CAT and the most relevant online assessment system with adaptive capabilities. Section 3 presents the first study regarding the students' opinions on FIT and CAT, whereas Sect. 4 delves into the different methods for converting the ability into a grade. Section 5 discusses and relates the results of the two studies, and Sect. 6 ends the paper with short conclusions and the future work.

2 Background

2.1 CTT/FIT, IRT/CAT

CTT and IRT are the two methods commonly employed to perform the analysis of test items. As mentioned in the introduction, according to CTT, the student learning is assessed through FIT as the unweighted sum of the answers to the test items, plus a casual error component [11]. The use of FIT during the exams, while remaining a valuable and largely used tool, presents some limitations [10,14]. For instance, by correctly answering questions of different difficulty, students get the same score. Furthermore, the error component is not invariant to the sample of test-takers and it is therefore not a characteristic of the test itself [27].

On the other hand, IRT [5] defines models based on parameters (e.g., difficulty, discrimination, guessing) and algorithms to measure the ability of an examinee. According to IRT, the ability is a numeric value, generally represented with the Greek letter theta (θ). Ability values are usually included in the range between $[-3, 3]$, although in theory they could range between $-\infty$ to ∞. According to IRT, it is possible to select specific items for each examinee – given the most current information about the examinee's proficiency and all possible items – and therefore each examinee can have a tailored test. Such a process can be automatised, thus leading to Computerised Adaptive Testing (CAT) [32]. Adaptation is a valuable approach to support learning [23] and it is not limited to the use of items, e.g., using students' models and artificial intelligence methods [1,12,16].

Focusing on items, a classical CAT algorithm (see Fig. 1) consists of some basic components: (i) an item selection criteria for the first item; (ii) the scoring process of the examinee's responses to obtain the estimated ability; (iii) an item selection criteria for other items; and (iv) the stopping criteria so to end test administration. In detail, see Fig. 1, a CAT system starts assuming that the examinee is of an initial ability. Then, it selects a set of initial items, usually with average difficulty. After the examinee's answers, the examinee's ability is estimated. Then, a new item is administered, so that it is either the most informative within the item bank, or gives the greatest weighted information, or determines the greatest reduction in the variance of the posterior distribution. These phases are repeated until a stopping criterion is reached, usually based on standard error estimation. So, unlike CTT/FIT, answering more questions correctly does not increase the overall ability estimate unless the items are hard. In other words, in CAT, the examinee's score increases, not by answering a higher proportion of questions correctly, but by correctly answering much harder questions [6].

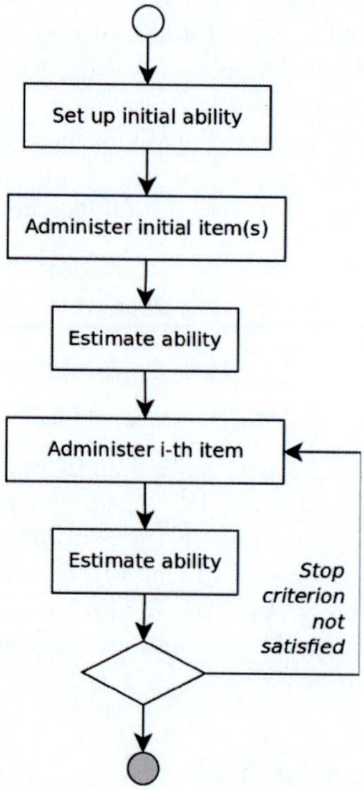

Fig. 1. Flowchart followed by a CAT system

Studies show that CAT can measure the ability of different examinees with the same level of precision, higher than that of classical tests, by also reducing the number of administered items (e.g., [2,20,30,32,33]). Furthermore, CAT seems to provide an appropriate challenge for each examinee, so that low examinees are not discouraged whereas high examinees may enjoy receiving challenging items, which should result in an increased engagement. However, this point is still debated and the results available in the literature are not univocal [2,21,29].

2.2 Online Assessment Systems

Several online systems for online assessment are available. Hereafter, we summarise the most relevant ones, focusing only on those offering computerised adaptive testing.

Moodle [3] is a complete e-learning platform, more than an online assessment system. Focusing on the assessment capabilities, the question types available in the Moodle quizzes are of different types: from automatically marked (multiple-choice, true/false, short answer and matching), to interactive (drag & drop), and essay responses with manual marking. Moodle is structured as an application core, with numerous plugins that provide specific functionality. Among the many, of interest for our research is the plugin developed within the SIETTE project [9], that enables CAT assessments. SIETTE is not only a Moodle plugin, but it is instead a domain-independent web-based system for automated assessment. SIETTE has been mainly used in higher education because it was developed in a University environment and tailored to the needs of assessment at this level of education.

The leading open-source platform specialised in adaptive testing is Concerto [28]. It permits users to create online assessments with textual and graphical feedback. In particular, Concerto uses R as the back-end for all calculations, i.e., to perform statistical estimations, update variables, and complete the other processes for the intended assessment [22,25]. It also provides a modern user interface to the end-users.

The last reviewed system is the UnivAQ Test Suite (UTS) [4], which we are currently using in our Institution. UTS permits teachers to create online assessments, based on CTT/FIT, IRT/CAT, but also to use state-of-the-art artificial intelligence models for the (semi-)automated correction of solutions to programming exercises [18,31]. Similarly to Concerto, its adaptive capabilities are provided via an R back-end [7,22,25].

3 Students' Opinions

Objective. The objective of this study is to understand the students' opinions about CAT and FIT, highlighting the differences and strengths of the two methods. In particular, we investigated their preference, the perceived fairness and efficiency, the required preparation, if the sequential way of posing questions increased their attention in answering, and their anxiety during the assessment.

Materials and Methods. To this aim, students of Medicine and Surgery at the University of L'Aquila (Italy) were asked to answer the questionnaire detailed in the Appendix. The involved students used FIT (through Moodle) and CAT (through UTS) for their exams, depending on the assessment method decided by the professor of the specific subject. At the time of the study, all students have experienced both FIT and CAT in their training and assessment activities.

In short, the questionnaire asks a student his/her (i) attitude towards technology, (ii) opinions about CAT and FIT (i.e., if in favour, if fair, if fast to complete, the amount of needed preparation), then (iii) sequential modality of CAT and (iv) perceived anxiety.

In detail, the questionnaire is composed of 11 items, three of which are generic questions (i.e., $Q_{1.1}$, $Q_{4.1}$ and $Q_{4.2}$), the other eight centred on the acceptability of the two tests. The latter ones are composed of two blocks of 4 items each ($Q_{2.1}$ – $Q_{2.4}$, $Q_{3.1}$ – $Q_{3.4}$). Each block proposes the same questions regarding the perception of the students with respect to CAT and FIT, respectively. Each question is answered with the following 5-points Likert scale: "1 = Completely disagree", "2 = Disagree", "3 = Unsure", "4 = Agree", "5 = Completely agree". For the two blocks, we calculated also two scores as the sum of all related answers. The answers to the items $Q_{2.4}$ "I think that CAT needs more preparation than FIT" for the CAT block, and the analogue $Q_{3.4}$ "I think that FIT needs more preparation than CAT" for the FIT block were inverted, because considered unfavourably. Accordingly, we defined the following two scores:

$$
\begin{aligned}
CAT_{sum} &= Q_{2,1} + Q_{2,2} + Q_{2,3} + (6 - Q_{2,4}) \\
FIT_{sum} &= Q_{3,1} + Q_{3,2} + Q_{3,3} + (6 - Q_{3,4})
\end{aligned}
$$

Each score ranges from 4 to 20. The higher the score, the better the student's opinion about the respective test type.

To analyse the data, we used frequency tables and means for descriptive statistics, and Wilcoxon rank-sum tests to test the differences between CAT and FIT [26]. We used a non-parametric test given the qualitative nature of the answers.

Results. A total of 204 students were invited to the survey, and 72 students participated in the survey (35%).

Figure 2 describes the results in terms of stacked bars.

The largest part of the students (i.e., 53.5%) agree to have a good attitude towards technology ($Q_{1.1}$).

With respect to the questions composing each block, the answers are placed side by side, so to simplify the comparison. The students were more in favour of CAT ($p < 0.001$), perceived CAT fairer than FIT ($p < 0.001$), and considered CAT more practical and fast than FIT ($p < 0.001$). On the other hand, no statistical difference exists in terms of needed preparation. In terms of the overall scores, the averages are $\overline{CAT}_{sum} = 16.1$ and $\overline{FIT}_{sum} = 13.4$. The difference is statistically significant ($p < 0.001$).

As for the question regarding if the sequential way of asking questions supports focusing on the right answer ($Q_{4.1}$), the 48.6% and 31.9% of the students answered in total agreement and in agreement, respectively.

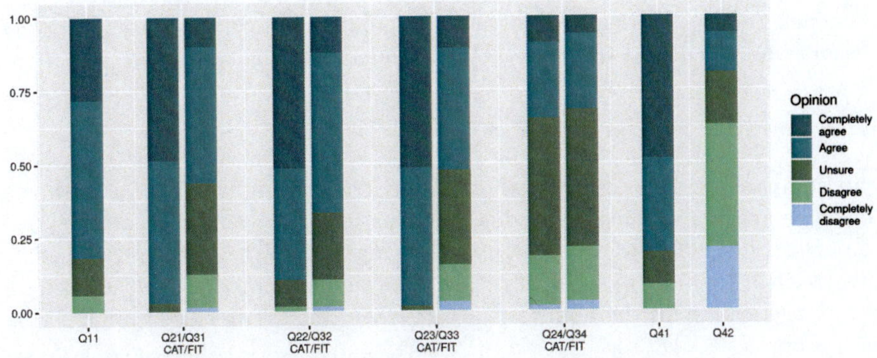

Fig. 2. Relative frequencies of answers, stacked bar

Moreover, only 13.4% of the students reported they were more anxious while attending a CAT test, instead of FIT. Conversely, 62.7% of the student completely disagree or disagree with this statement.

Finally, the different attitude toward technology does not affect – on average, in terms of statistical significance – any answer.

4 Ability Conversion Methods

Objective. Given that the final aim of the assessment is to provide the most accurate grade to a student, we discuss several methods to convert the ability into a grade and their effects in a real setting.

Material and Methods. Grading in the Italian system is a number ranging from 0 to 30, potentially "cum laude", customarily considered as 31. On the other hand, as discussed in Sect. 2.1, the ability measured through a CAT is a real number usually in the range $[-3, 3]$. Accordingly, so far, in the exams, the ability was converted automatically to the corresponding grade (let us call it γ) using the straightforward formula (linear conversion: $\theta = -3 \Rightarrow \gamma = 0$, $\theta = 3 \Rightarrow \gamma = 31$):

$$\gamma = 31 \cdot \frac{\theta + 3}{6}$$

Furthermore, any grade below zero was set to zero; any grade higher than 31 was set to 31.

To pursue the research objective, we proceeded as follows:

1. we collected the results of all CAT assessments, performed during the first session of the Academic year 2021/2022 (February 2022);

2. for each CAT assessment, we extracted the number of correct/wrong answers n_c, n_w and calculated the hypothetical grade according to CTT

$$\gamma_c = \frac{n_c}{n_c + n_w}$$

3. we compared the average hypothetical grade according to CTT $\overline{\gamma}_c$, with the average grade $\overline{\gamma}$ obtained with the following linear conversion strategies:
   (a) from $\theta \in [-3, 3]$ to $\gamma \in [0, 31]$: the original approach already discussed;
   (b) from $\theta \in [-4, 4]$ to $\gamma \in [0, 31]$: like the original approach, but adopting a larger range for the abilities;
   (c) from $\theta \in [\theta_{min}, \theta_{max}]$ to $\gamma \in [\gamma_{c,min}, \gamma_{c,max}]$, where θ_{min}, θ_{max}, $\gamma_{c,min}$, $\gamma_{c,max}$ are the minimum/maximum ability/grade as experimentally observed during the examinations: the idea is to map the range of ability observed during the exams to the range of possible grading as of CTT.

Results. A total of 265 exams were analysed. The analyses show the following. If we had used CTT/FIT, we would have had an $\overline{\gamma}_c = 26.6$. With the IRT/CAT approach, the different conversion strategies would have lead to: (a) $\overline{\gamma} = 28.4$, (b) $\overline{\gamma} = 27.0$ (c) $\overline{\gamma} = 24.8$.

These results show that using the common range for the ability, i.e., option (a), yields higher grades (on average) than with CTT (1.8 points more). The larger range for the ability, i.e., option (b), reduces this difference (0.4 points more). Finally, tailoring the conversion to the observed range, i.e., option (c), seems a penalising choice (1.8 points less, on average)[1].

5 Discussion

The results of the first study – in summary – suggest a preference for CAT assessment, considered fairer, more practical and faster than FIT, independently of the different attitudes towards technology. Moreover, referring to the necessary preparation, the students did not manifest any difference. In connection with the last two generic questions, most of the students considered the sequential mode in asking questions (which is specific for CAT and optional for FIT) to support their concentration and disagree with the hypothesis to be more anxious in sustaining the exam with the CAT method. These results are in line with the literature, with the exception of perceived test anxiety.

The results concerning the different ability conversion methods show that the adoption of the larger range returns – on average – grades more similar to CTT, whereas the currently adopted method returns higher grades than CTT. Therefore, given that the adoption of the larger range seems to be fairer than the currently adopted method, and given that it is not ethical to change the

[1] For completeness, the range for θ that nullify the average difference between grades calculated through CTT and the proposed linear conversion is $[-4.15, 4.15]$.

conversion method for the same cohort of students, we planned to switch to the larger range in the next academic year. A second consideration is also a possible limitation in our study. Even if the hypothetical grade calculated with CTT is not known to students, it is possible that the students felt more "rewarded" by CAT, and this could have influenced their opinions.

As further limitations of this study, we mention the small cohort of students and that two different systems were used for CAT and FIT (UTS and Moodle, respectively). Furthermore, to measure the anxiety, instead of validated and standardised questionnaires adopted in the studies mentioned in Sect. 2.1, we only used a single question.

6 Conclusions

The moment of the final assessment, aimed at measuring the level of the objectives obtained in relation to the knowledge acquired of the considered topics, is always critical. The maximisation of the reliability and precision of the obtained measurements, to better reflect the degree of reached competencies, is both for the teacher and student the last and most important goal. This consideration pushed the authors to use CAT assessments in place of FIT.

Starting from the scientific literature, in this research we focused on evaluating the acceptability and opinions of students towards CAT. Despite some limitations of the study, we collected useful information regarding the preference of students toward CAT and how to improve the fairness of the ability-to-grade conversion. Moreover, students did not report increased anxiety in using CAT or additional workload.

If previous research showed CAT faster, more precise in estimating the student's ability, more engaging than FIT [2,21,30,33], in this experience we did not observe an increased anxiety (as reported e.g. in [8,17]). Moreover, students were in favour of CAT and their opinions followed the literature. Accordingly, the lesson learned from this study is for teachers interested in the use of adaptive testing: in our experience, CAT was positively perceived by students as a valuable tool for the assessment and could be used as an efficient and fair substitute for FIT.

Questionnaire

		1	2	3	4	5
Personal attitude						
$Q_{1.1}$	My attitude towards technology is good					
Computerised Adaptive Test (CAT)						
$Q_{2.1}$	I am in favour of using CAT					
$Q_{2.2}$	I perceived CAT as fair					
$Q_{2.3}$	CAT is practical and fast					
$Q_{2.4}$	I think that CAT needs more preparation than FIT					
Fixed-Item Test (FIT)						
$Q_{3.1}$	I am in favour of using FIT					
$Q_{3.2}$	I perceived FIT as fair					
$Q_{3.3}$	FIT is practical and fast					
$Q_{3.4}$	I think that FIT needs more preparation than CAT					
General						
$Q_{4.1}$	The sequential way of asking questions allowed me to focus more on the right answer					
$Q_{4.2}$	Regardless of the subject, taking an exam with CAT, rather than FIT, made me feel more anxious					

Each question is answered with the following 5-points Likert scale: "1 = Completely disagree", "2 = Disagree", "3 = Unsure", "4 = Agree", "5 = Completely agree".

References

1. Alrifai, M., Gennari, R., Vittorini, P.: Adapting with evidence: the adaptive model and the stimulation plan of TERENCE. In: Vittorini, P., Gennari, R., Marenzi, I., de la Prieta, F., Rodríguez, J. (eds.) International Workshop on Evidence-Based Technology Enhanced Learning. AISC, vol. 152, pp. 75–82. Springer, Heidelberg (2012). https://doi.org/10.1007/978-3-642-28801-2_9
2. Angelone, A.M., Vittorini, P.: A report on the application of adaptive testing in a first year university course. In: Uden, L., Liberona, D., Sanchez, G., Rodríguez-González, S. (eds.) LTEC 2019. CCIS, vol. 1011, pp. 439–449. Springer, Cham (2019). https://doi.org/10.1007/978-3-030-20798-4_38
3. Athaya, H., Dwi, R., Nadir, A., Sensuse, D.I., Suryono, R.R.: Moodle implementation for e-learning: a systematic review. In: 6th International Conference on Sustainable Information Engineering and Technology 2021, vol. 14 (2021). https://doi.org/10.1145/3479645.3479646
4. Bernardi, A., et al.: On the design and development of an assessment system with adaptive capabilities. In: Di Mascio, T., et al. (eds.) MIS4TEL 2018. AISC, vol. 804, pp. 190–199. Springer, Cham (2019). https://doi.org/10.1007/978-3-319-98872-6_23

5. Birnbaum, A.: Some latent trait models and their use in inferring an examinee's ability. In: Statistical Theories of Mental Test Scores, pp. 395–479 (1968)
6. Brown, G.T.L., Abdulnabi, H.H.A.: Evaluating the quality of higher education instructor-constructed multiple-choice tests: impact on student grades. Front. Educ. **2**, 24 (2017). https://doi.org/10.3389/feduc.2017.00024
7. Chalmers, R.P.: mirt: a multidimensional item response theory package for the R environment. J. Stat. Softw. **48**(6), 1–29 (2012). https://doi.org/10.18637/jss.v048.i06
8. Colwell, N.M.: Test anxiety, computer -adaptive testing, and the common core. J. Educ. Train. Stud. **1**(2), 50–60 (2013). https://doi.org/10.11114/JETS.V1I2.101
9. Conejo, R., Guzmán, E., Trella, M.: The SIETTE automatic assessment environment. Int. J. Artif. Intell. Educ. **26**(1), 270–292 (2016). https://doi.org/10.1007/s40593-015-0078-4
10. De Champlain, A.F.: A primer on classical test theory and item response theory for assessments in medical education. Med. Educ. **44**(1), 109–117 (2010). https://doi.org/10.1111/j.1365-2923.2009.03425.x
11. DeVellis, R.F.: Classical test theory. Med. Care **44**(11), S50–S59 (2006). https://doi.org/10.2307/41219505
12. Di Mascio, T., Gennari, R., Melonio, A., Vittorini, P.: The user classes building process in a TEL project. In: Vittorini, P., Gennari, R., Marenzi, I., de la Prieta, F., Rodríguez, J. (eds.) International Workshop on Evidence-Based Technology Enhanced Learning. AISC, vol. 152, pp. 107–114. Springer, Heidelberg (2012). https://doi.org/10.1007/978-3-642-28801-2_13
13. Dunn, L., Morgan, C., O'Reilly, M., Parry, S.: The Student Assessment Handbook. Routledge, Abingdon (2003). https://doi.org/10.4324/9780203416518
14. Eleje, L.I., Onah, F.E., Abanobi, C.C.: Comparative study of classical test theory and item response theory using diagnostic quantitative economics skill test item analysis results. Eur. J. Educ. Soc. Sci. **3**(1), 57–75 (2018)
15. Embretson, S.E., Reise, S.P.: Item Response Theory for Psychologists. Lawrence Erlbaum Associates, Mahwah (2000)
16. Erdt, M., Fernández, A., Rensing, C.: Evaluating recommender systems for technology enhanced learning: a quantitative survey. IEEE Trans. Learn. Technol. **8**(4), 326–344 (2015). https://doi.org/10.1109/TLT.2015.2438867
17. Fritts, B.E., Marszalek, J.M.: Computerized adaptive testing, anxiety levels, and gender differences. Soc. Psychol. Educ. **13**(3), 441–458 (2010). https://doi.org/10.1007/S11218-010-9113-3
18. Galassi, A., Vittorini, P.: Automated feedback to students in data science assignments: improved implementation and results. In: 14th Biannual Conference of the Italian SIGCHI Chapter (CHItaly 2021), New York, NY, USA. ACM, Bolzano (2021). https://doi.org/10.1145/3464385.3464387
19. Gikandi, J., Morrow, D., Davis, N.: Online formative assessment in higher education: a review of the literature. Comput. Educ. **57**(4), 2333–2351 (2011). https://doi.org/10.1016/J.COMPEDU.2011.06.004
20. Hambleton, R.K., Jones, R.W.: Comparison of classical test theory and item response theory and their applications to test development. Educ. Meas. Issues Pract. **12**(3), 38–47 (1993). https://doi.org/10.1111/j.1745-3992.1993.tb00543.x
21. Ling, G., Attali, Y., Finn, B., Stone, E.A.: Is a computerized adaptive test more motivating than a fixed-item test? Appl. Psychol. Meas. **41**(7), 495–511 (2017). https://doi.org/10.1177/0146621617707556

22. Magis, D., Barrada, J.R.: Computerized adaptive testing with R: recent updates of the package catR. J. Stat. Softw. **76**(Code Snippet 1), 1–19 (2017). https://doi.org/10.18637/jss.v076.c01
23. Mulwa, C., Lawless, S., Sharp, M., Arnedillo-Sanchez, I., Wade, V.: Adaptive educational hypermedia systems in technology enhanced learning: a literature review. In: Proceedings of the 2010 ACM Conference on Information Technology Education - SIGITE 2010 (2010). https://doi.org/10.1145/1867651
24. Qasem, M.A.N.: A comparative study of classical theory (CT) and item response theory (IRT) in relation to various approaches of evaluating the validity and reliability of research tools. IOSR J. Res. Method Educ. (IOSR-JRME) **3**(5), 77–81 (2013)
25. R Core Team: R: A Language and Environment for Statistical Computing (2018). https://www.R-project.org/
26. Riffenburgh, R.H.: Statistics in Medicine. Elsevier/Academic Press, Cambridge (2012)
27. Rindskopf, D.: Reliability: measurement. Int. Encycl. Soc. Behav. Sci. 13023–13028 (2001). https://doi.org/10.1016/B0-08-043076-7/00722-1
28. Scalise, K., Allen, D.D.: Use of open-source software for adaptive measurement: concerto as an R-based computer adaptive development and delivery platform. Br. J. Math. Stat. Psychol. **68**(3), 478–496 (2015). https://doi.org/10.1111/BMSP.12057
29. Tonidanel, S., Quinones, M.A.: Psychological reactions to adaptive testing. Int. J. Sel. Assess. **8**(1), 7–15 (2000). https://doi.org/10.1111/1468-2389.00126
30. Vispoel, W.P., Rocklin, T.R., Wang, T.: Individual differences and test administration procedures: a comparison of fixed-item, computerized-adaptive, and self-adapted testing. Appl. Measur. Educ. **7**(1), 53–79 (1994). https://doi.org/10.1207/s15324818ame0701_5
31. Vittorini, P., Menini, S., Tonelli, S.: An AI-based system for formative and summative assessment in data science courses. Int. J. Artif. Intell. Educ. 1–27 (2020). https://doi.org/10.1007/s40593-020-00230-2
32. Wainer, H., et al.: Computerized Adaptive Testing. Routledge, Abingdon (2000). https://doi.org/10.4324/9781410605931
33. Weiss, D.J., Kingsbury, G.G.: Application of computerized adaptive testing to educational problems. J. Educ. Measur. **21**(4), 361–375 (1984). https://doi.org/10.1111/j.1745-3984.1984.tb01040.x

Readability Assessment of Academic Texts at Different Degree Levels

José Medardo Tapia-Téllez[1], Aurelio López-López[1](✉), Samuel González-López[2], and Jesús Miguel García-Gorrostieta[3]

[1] Instituto Nacional de Astrofísica Óptica y Electrónica, Puebla 72840, Mexico
allopez@inaoep.mx

[2] Instituto Tecnológico de Nogales, Nogales 84065, Mexico
samuel.gl@nogales.tecnm.mx

[3] Universidad de la Sierra, Sonora, Mexico
jgarcia@unisierra.edu.mx

Abstract. Developing machine learning tools to aid students in the process of writing a thesis document is of great interest to students, universities, supervisors and evaluation committees. This article presents the construction and evaluation of readability comparators based in Spanish-written thesis documents of four different academic levels: Advanced College Level Technician (ACT), Undergraduate, Master and Doctoral. Specifically, we provide comparators that can evaluate, between two academic texts which one is more readable than the other. In particular, we focus on three thesis sections: Problem Statement, Justification, and Results. The successful completion of these different comparators, as shown in results, allowed to test readability evaluators for new undergrad writings, determining whether correspond to its academic level. Some guidelines to determine comparators and academic level archetypes for readability assessment are proposed, based on our results.

Keywords: Readability assessment · Academic texts · Text sorting · Undergraduate text evaluation · Machine learning

1 Introduction

The quest for aiding students in the production of a thesis document is a longtime problem that affects universities, thesis supervisors, and committees. According to [1], students in general demand flexible forms and structures of research that can explode the extensive use of new technologies. Our aim is to provide the basis to develop these new technologies, specifically for writers in Spanish.

Computational-linguistic technologies such as word correctors or grammar checkers have aided students in the production of a thesis document. We now face a Machine Learning revolution, where tools such as Grammarly[1] assists students and public in general in writing documents of any kind. Such tools are

[1] https://www.grammarly.com.

M. Temperini et al. (Eds.): MIS4TEL 2022, LNNS 580, pp. 37–47, 2023.
https://doi.org/10.1007/978-3-031-20617-7_6

very helpful but have some limitations: first, they were developed for English, and secondly, they are specifically focused on grammar. We need to develop, first these kind of tools for Spanish, and then, we must include deeper linguistic analysis in them. In this work, we report the first steps to reach a readability evaluator of thesis.

According to [2], text readability is the sum of all elements in textual material that affect a reader's understanding, reading speed, and level of interest in the material.

The main contributions of this work are: a) the construction of a set of classifiers (readability comparators) for two academic levels; b) the application of the classifiers to assess the readability of a new academic text, and c) the determination of the best representative (archetype) of each academic level for the evaluation of undergraduate writings.

The document is organized as follows. Section 2 details related work to our research. Section 3 presents the text collection for experiments, along with its statistics. Section 4 describes the methodology for readability evaluation, comparators construction, and evaluator process. We conclude in Sect. 5 with future directions.

2 Related Work

To obtain an extensive background on how text readability is assessed automatically we found [3]. They review the state-of-the-art approaches in modeling and predicting the reading difficulty of texts, and also list new challenges and opportunities in the area. In general, studies on this area fall within regression and classification, however, one can find research that treats this task as a pairwise problem. This work allowed to point out the not so explored idea of using comparators as a machine learning tool to evaluate readability of documents.

The work of [4] successfully builds a comparator through machine learning that can judge text readability between two texts. They then utilize this comparator to sort a set of texts and present an application which retrieves texts with readability similar to that of a given input text. Here, we employ this comparator construction scheme but we implement it for Spanish-written theses, focusing on specific sections.

To build a comparator, a document representation suited for our type of documents is needed. In [5], they combine lexical, syntactic, and discourse features to produce a highly predictive model of human reader's judgments of text readability. This is a work that treats readability prediction as a pairwise preference learning problem, thus estimating the relative difficulty of pairs of documents instead of assigning a specific level. Through this work, we could obtain state-of-the-art feature representation for our comparator.

Finally, since our interest is to use the comparators as tools to support students in the process of thesis writing, [6] is relevant. They evaluate the quality of eBooks through text analytics and identify parts that need improvement. In [7], they theorize about the generation of an automatic validation for a text that contains information pertaining to a medical procedure, and give future approaches on why these tools are viable and an important subject of research.

3 Document Collection

For our research, we use the document collection of [8]. The data set consists of theses and proposals of different academic levels (previously reviewed and approved by a committee), such as: Advance College-level Technician[2] (ACT), Undergraduate, Master, and Doctoral. The following sections were extracted from each document: Problem Statement, Justification, and Results. Table 1 includes the number of theses of each level in the collection (All), the selected theses (i.e. documents with more than one paragraph), and the number of sections of interest.

Table 1. Collection data by academic level.

	All	Selected	Problem Stmt	Justification	Results
ACT	227	202	80	104	102
Undergraduate (UG)	150	136	28	21	77
Master	269	254	92	81	179
Doctoral	66	64	21	13	43

Word/token statistics for the three sections in the document collection is given in Table 2, that includes average, larger text (Max) and shorter text (Min).

Table 2. Statistics of tokens per section and academic level.

	Problem Stmt			Justification			Results		
	Avg.	max.	min.	Avg.	max.	min.	Avg.	max.	min.
ACT	409.33	1708	75	385.07	1123	110	493.55	1883	138
UG	477.60	1253	114	342.33	820	109	508.92	2386	121
Master	399.39	1063	121	407.81	1612	113	656.65	5184	137
Doctoral	754.85	1643	310	433.84	802	247	840.27	3765	59

2 A two year program offered in some countries.

4 Readability Evaluation Approach

The proposed approach to assess readability of academic texts consists of two phases. First a set of comparators (classifiers) are trained, where each comparator receives as examples pairs of documents of two different academic levels. Then, in the second phase, having the classifiers between the different levels, a new document (possibly a draft or document in progress) is submitted for evaluation to verify whether it corresponds to the intended academic level. Our aim is to support undergraduate learners. Next, these two phases are detailed.

4.1 Comparator Construction

Each comparator is trained through machine learning and can judge, given two texts, which one is more readable than the other. To build the comparator, we have to first determine a representation for documents and then train the model.

Document Representation. As detailed in Sect. 3, we extracted sections of interest from each document (thesis). To get a vector to feed in the machine learning model, we must have two texts from different theses, each of the same kind of section but different academic levels. Let call these texts $a, b \in S$, with S the set of texts selected from all the thesis documents. To build the feature vectors V_a and V_b, we extract local and global characteristics from a and b, as shown in Fig. 1, where each V_a and V_b has its local and global features. By local characteristics, we get the frequency of each word divided by the frequency of the number of words in the text, i.e. the relative frequency. By global features, we refer to the log frequency of the 5000 most common words in Spanish[3].

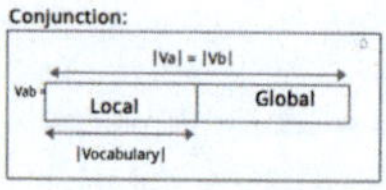

Fig. 1. Feature representation for text documents.

Then, a single vector from vectors V_a and V_b is obtained by operate them. In Fig. 1, we can observe that the conjunction operation of V_a with V_b produces a V_{ab} vector of the same size, since this considers two possible operations: vector difference and vector division. Even though we experimented with both vector operations, difference worked better.

[3] [9] is a dataset created by the Royal Academy of (Spanish) Language, that provides statistics of the most common words.

Training the Comparator. V_{ab} vectors come from a set of sections of two different academic levels, let call them A and B, with $a \in A$, $b \in B$, $|A| = n$ and $|B| = m$. Each vector of a document in A is operated with each of the vectors of documents in B and classified as $+1$ if A is of a higher academic level than B and -1 otherwise, leading to $n \times m$ vectors V_{ab}. We also perform this procedure the other way around, i.e. having B first, and thus getting $m \times n$ vectors V_{ba}. This is done since V_{ab} vectors are not the same as V_{ba} vectors and their classification differs. So, we end up with a matrix of size $2 \times n \times m$ for training.

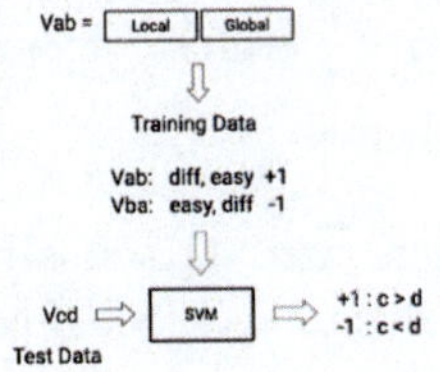

Fig. 2. Training and testing diagram for comparator.

This matrix is used to train the selected machine learning models for our comparators, as illustrated in Fig. 2, where vectors of the form V_{ab} and V_{ba} are fed into a Support Vector Machine (SVM). After training, Fig. 2 also depicts a V_{cd} vector input to the machine, that was created from two texts: c and d. These texts were vectorized and operated, leading to the V_{cd} vector which can be fed to the trained SVM to get an evaluation of their relative readability.

Comparator Experiments. Given that texts are of four different academic levels, and each of the comparators is built from two different academic levels, we can have up to six: ACT-Undergrad, ACT-Master, ACT-Doctoral, Undergrad-Master, Undergrad-Doctoral, and Master-Doctoral. Likewise, for each thesis document, we extracted three sections of interest: Problem Statement, Justification, and Results. Three different machine learning models were applied: SVM, K-Nearest Neighbors (K-NN), and Perceptron.

For the experiments, the selected sections detailed in Table 1 were randomly divided in 80% for training and 20% for test. For instance, 83 justifications of ACT level were combined with 64 of master level, leading to more than 10K training instances. We evaluate the classifier efficacy in terms of accuracy given that the data set is balanced (has equal number of $+1$ and -1 instances).

Results of Comparators. In Table 3, we can observe that the comparator with the worst efficacy is ACT-Undergraduate across all sections. The best result corresponds to Undergraduate-Doctoral comparator followed by ACT-Doctoral. Regarding average, SVM produces the best averages across Problem Statement and Justification, and Perceptron slightly better in Results. K-NN did not compete. So, SVM showed a more consistent efficacy.

Table 3. Accuracy percentage for comparators trained with vector operated by difference on three different machine learning models across all sections.

Comparator	Problem Stmt			Justification			Results		
	SVM	KNN	Perceptron	SVM	KNN	Perceptron	SVM	KNN	Perceptron
ACT-Undergraduate	36.65	41.59	37.39	64.70	53.61	63.87	46.38	56.26	42.41
ACT-Master	54.41	55.55	53.18	61.99	64.93	56.97	64.14	62.17	69.43
ACT-Doctoral	81.09	55.46	82.14	83.58	67.69	73.33	82.14	65.87	81.15
Undergraduate-Master	57.53	47.42	47.81	70.84	50.34	65.31	77.67	**74.96**	81.73
Undergraduate-Doctoral	**90.58**	**77.64**	**87.64**	**90.58**	**77.64**	**87.64**	**86.11**	62.64	**85.64**
Master-Doctoral	77.64	67.64	64.11	77.64	67.64	64.11	74.35	53.20	75.40
Average	66.31	57.55	62.04	74.88	63.64	68.53	71.79	62.51	72.62

Comparators Result Analysis and Discussion. We notice that comparators trained with more distant academic levels obtain higher classification accuracy values, e.g. ACT-Undergraduate is lower on accuracy than ACT-Doctoral. This one being an important observation, since it validates our results and provides insight into the learning capabilities of the machine learning models.

4.2 Evaluators Process

Once the classifiers are trained, the second phase and main interest for readability assessment is the development of evaluators, capable of judging the academic level of a submitted text. The issue to resolve, given the input text, is determine an archetype of academic level to compare the text. For this purpose, four representatives are explored: Minimum, Maximum, Centroid, and Random.

Figure 3 illustrates the process to assess whether a given document is more or less readable than one of a given academic level. The input text is expressed in the same representation (a vector) as that used to train the classifiers. Also an archetype vector of a given academic level is taken (either Minimum, Maximum, Centroid, or Random). Then both vectors are operated and submitted to the classifier, that decides which is more readable than the other.

Maximum and minimum representatives of an academic level are obtained through the use of the classifier, selecting the first and second vector of the collection as maximum or minimum (depending on whether the first or the second is more complex than the other). This selection is compared to all the other vectors in the collection; if anyone turns out to be more difficult or less complex (according to the comparator), then is taken as new maximum (or minimum). Centroid and random representatives are directly computed or taken stochastically.

With this information, we can bound the academic level readability of the document under evaluation. For instance, if we introduce a document of undergrad level and we compare it against an archetype of master level, we would expect it to be of a lower readability level, contrary to a ACT level where we would expect it of a higher readability.

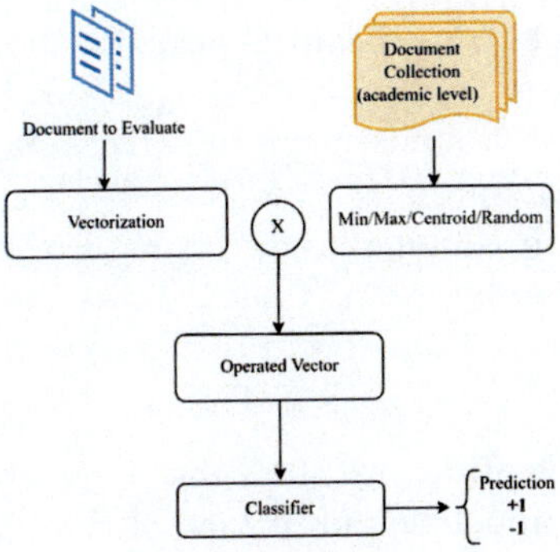

Fig. 3. General scheme for evaluators.

Evaluator Experiments. Our experiments focus on the evaluation of undergraduate (bachelor) level documents for two reasons: first, these documents are at an academic level that is between two other academic levels of study; second, we plan to use these evaluators in a web application to help learners to improve their writings. The first stage would be for students pursuing a bachelor's degree.

For this purpose, we then employ the 20% of undergraduate level documents to evaluate the different archetypes. These texts were not used for training.

To get an upper bound, the comparators that have been trained with higher academic levels are selected, and these comparators must have the best efficacy of Table 3. The comparators selected to experiment were Undergraduate-Master and Undergraduate-Doctoral, the latter because its classification stood out notably from the other. The machine learning model used was based on that with the best accuracy for a respective comparator, so it could vary between SVM, K-NN, and Perceptron. In each table of results of experiments, it is clarified which of these was explicitly taken. Regarding the selection of comparators for lower bounds, those at the ACT-Master level were selected due to their superior results compared to ACT-Undergraduate level, as Table 3.

The results of the experiments for the evaluators are presented next. In each of the tables, the archetypes of interest are in headers, i.e. whether results with upper bounds are being studied, we are interested in results with representatives of higher level than the academic level of the document to be evaluated. We study evaluators with comparators that should intuitively be better for upper bounding (Undergraduate-Master and Undergraduate-Doctoral), but we also study evaluators for lower bounding (ACT-Undergraduate). Hence, we are interested in observing how the archetypes of Doctoral and Master level work for upper bounds. When experimenting with lower bound, we are interested in finding the behavior of the representative of ACT level. Through these experiments, we will be able to select the best evaluator (archetype in combination with its comparator) to delimit both higher and lower document levels.

Problem Statement Section Evaluators. In the results of the Table 4, we observe that both master and doctoral archetypes behave as expected, but the doctoral representative bounds more effectively. Regarding the evaluators, the worst evaluator to generate an upper bound is the Minimum, followed by the Centroid. Finally, both Random and Maximum show to be effective upper bounds and either one could be taken. Table 5 shows that the archetypes of ACT level work well to lower-bound undergrad texts, except by the Maximum.

Justification Section Evaluators. In Table 6, we observe that the doctoral archetype is an excellent upper bound in each of the evaluators. Regarding the master representative, we observe that is an excellent upper bound with the Maximum evaluator and moderately good with the Random and Centroid evaluators. However, Minimum does work at all. Table 7 shows the results that we hypothesized since the archetype of Advance College-level Technician is the best lower bound we have for the Justification section, with Centroid not working as expected.

Results Section Evaluators. In Table 8, we observed that the Doctoral level is an effective upper bound with all its evaluators. The Master level is an effective upper bound for the Minimum and Maximum evaluators. In Table 9, we notice that the ACT level works well as lower bound for this section and, in particular, its Minimum evaluator is the most effective, followed by Random and Centroid.

Evaluators Analysis and Discussion. As observed from the experiments with evaluators for the three sections, the best way to upper bound an undergrad text under evaluation is employing the Undergraduate-Doctoral comparator, working well with any archetype. However, if we are constrained to the next academic level (Master), the representative that worked consistently well was Maximum, or Centroid otherwise. Random or Minimum does not seem to work in this case.

In terms of lower bounding an undergrad text, Minimum is the best choice, followed by a Random or Centroid archetype. Maximum would be the worst choice for this purpose.

The diagnostic of a draft would be that satisfies or overpasses the readability for its academic level. But other possibility is that is below its level and requires further work. In this case, some suggestions can be given such as: improve or vary the vocabulary.

5 Conclusion

The present work successfully built and evaluated machine learning based readability comparators trained with Spanish-written thesis documents of four different academic levels. These comparators can decide between two different thesis sections, which one is more readable. In general, results across all sections showed that the best comparator was Undergraduate-Doctoral, the most stable model to train them was SVM with a representation using difference operator.

We further tested the trained classifiers (comparators) to assess new texts at undergrad level, working well to bound them properly. Some guidelines to perform this evaluation were reached after experiments.

We will also explore the possibility of incorporating a pairwise ranking model as that of [10], that brings deep learning in our research. Finally, we are interested in deploying our machine learning-based comparator in a web-based API that, given an input document, determines its readability level.

Table 4. Results in accuracy for Problem Statement evaluators with different archetypes (SVM).

Archetype	Random	Centroid	Min	Max
Master	100%	76%	0%	100%
Doctoral	100%	100%	100%	100%

Table 5. Results in accuracy of Problem Statement evaluators with different archetypes (K-NN).

Archetype	Random	Centroid	Min	Max
ACT	100%	100%	100%	15%

Table 6. Results in accuracy for Justification evaluators with different archetypes (SVM).

Archetype	Random	Centroid	Min	Max
Master	65%	65%	0%	100%
Doctoral	100%	100%	100%	100%

Table 7. Results in accuracy of Justification evaluators with different archetypes (K-NN).

Archetype	Random	Centroid	Min	Max
ACT	100%	88%	100%	100%

Table 8. Results in accuracy for evaluators with different archetypes (SVM).

Archetype	Random	Centroid	Min	Max
Master	51%	74%	100%	100%
Doctoral	100%	100%	100%	100%

Table 9. Results in accuracy of evaluators with different archetypes (Perceptron).

Archetype	Random	Centroid	Min	Max
ACT	81%	81%	100%	33%

Acknowledgements. The first author thanks the support provided through scholarship 1009285 by Conacyt, México. The other authors were partially supported by SNI, México.

References

1. De-Anglat, H.D.: Las funciones del tutor de la tesis doctoral en educación. Rev. Mex. Investig. Educ. **50**(16), 935–959 (2011)
2. Chall, J.S., Dale, E.: Readability revisited: the new Dale-Chall readability formula. Brookline Books, Universidad de Michigan (1995)
3. Collins-Thompson, K.: Computational assessment of text readability: a survey of current and future research. ITL-Int. J. Appl. Linguist. **2**(165), 97–135 (2014)
4. Tanaka-Ishii, K., Tezuka, S., Terada, H.: Sorting texts by readability. Comput. Linguist. **2**(36), 203–227 (2010)
5. Pitler, E., Nenkova, A.: Revisiting readability: a unified framework for predicting text quality. In: Proceedings of the 2008 Conference on Empirical Methods in Natural Language Processing, EMNLP 2008, pp. 186–195. Association for Computational Linguistics (2008)
6. Relan, M., Khurana, S., Singh, V.K.: Revisiting readability: a unified framework for predicting text quality. In: Proceedings of the Seventeenth International Conference on Artificial Intelligence and Law, ICSES 2013 (2013)
7. Glaser, I., Bonczek, G.,Landthaler, J., Matthes, F.:Towards computer-aided analysis of readability and comprehensibility of patient information in the context of clinical research projects. In: Proceedings of the Seventeenth International Conference on Artificial Intelligence and Law, ICAIL 2019, pp. 260–261. Association for Computing Machinery (2019). https://doi.org/10.1145/3322640.3326704

8. González-López, S., López-López, A.: Colección de tesis y propuesta de investigación en tics: un recurso para su análisis y estudio. In: XIII Congreso Nacional de Investigación Educativa (2015)
9. Sánchez, M., Domínguez, C.: El banco de datos de la RAE: CREA y CORDE. Per Abbat: boletín filológico de actualización académica y didáctica **2**, 137–148 (2007)
10. Wang, L., Shen, X., De-Melo, G., Weikum, G.: Cross-domain learning for classifying propaganda in online contents. CoRR (2016)

Materials Science and Engineering Education Based on Reality-Virtuality Technologies

J. Extremera[1](✉), D. Vergara[2], and S. Rodríguez[1]

[1] University of Salamanca, 37008 Salamanca, Spain
{jamil.extremera,srg}@usal.es
[2] Catholic University of Ávila, 05005 Avila, Spain
diego.vergara@ucavila.es

Abstract. The use of reality-virtuality technologies (RVTs) in the field of engineering education has shown a growing trend for more than a decade. The teaching of Materials Science and Engineering (MSE) is no stranger to this phenomenon, and there are numerous examples of it. In order to better understand how RVTs are implemented in MSE teaching, this review presents an analysis that has been carried out after a systematic search in three different databases. This systematic search aimed to find academic works describing tools based on RVTs to improve the teaching-learning process of MSE. The results obtained provide an overall picture of how RVTs have been used since 2010 in teaching MSE: (i) virtual reality (immersive and non-immersive) is the most widely used technology, followed at a considerable distance by augmented reality, while mixed reality is a technology that is rarely used currently; (ii) the field of MSE on which most papers have focused is *Materials structure, processing, and properties*. Based on this information, it is possible to open up lines of research that have not been explored until now.

Keywords: Materials science and engineering · Virtual reality · Augmented reality · Mixed reality

1 Introduction

Recent times have seen a great development and expansion of Information and Communication Technologies (ICTs). Proof of this expansion can be seen in the large number of academic papers focusing on branches of ICT that the general public had barely heard of just two decades ago, such as machine learning [1–6], blockchain [7–13] and the internet of things [14, 15], among others. Hospital care [16], the design of future smart cities [17] or energy optimization in homes [18] are examples of how these technologies can help improve people's lives. Reality-virtuality technologies (RVTs) have undergone great development in parallel with these ICTs. The main RVTs that can be found on the market today are virtual reality (VR), augmented reality (AR) and mixed reality (MR). VR allows users to interact with computer-generated 3D environments [19], in a virtual world beyond the real one [20]. It is common to differentiate immersive VR

M. Temperini et al. (Eds.): MIS4TEL 2022, LNNS 580, pp. 48–58, 2023.
https://doi.org/10.1007/978-3-031-20617-7_7

(IVR, which uses some form of head-mounted display or CAVE-type system [21] to immerse the user in the virtual world) from non-immersive VR [22, 23] (NIVR, which displays virtual environments through a screen). Augmented reality (AR) allows the user to see the real environment "augmented" by superimposing computer-generated virtual elements above it [24]. Mixed reality (MR) also superimposes virtual elements on the real world but is more integrated and coherent with the real world than AR [25, 26].

For more than a decade, there has been an increasing trend in the use of RVTs in the field of engineering, as can be seen from the growing number of academic papers published annually since the year 2000 related to these technologies [27–29]. Technology is often used to facilitate the understanding of academic subjects through different techniques [30]. In this sense, there are numerous examples of the use of tools based on RVTs applied to improve the teaching-learning process of Materials Science and Engineering (MSE) [27, 31, 32], but it can be difficult for the researcher to know which areas are still to be explored in order to define new lines of research. This is why it is of great interest to have a "snapshot" that allows to quantify the way in which RVTs have been implemented to support the teaching of MSE. In particular, this work seeks to answer these two research questions:

- Question 1: in which areas of MSE do academic works that use RVTs to facilitate its learning focus on and in which measure?
- Question 2: which RVTs are used to facilitate the teaching of MSE and in what measure?

In this work, a systematic search has been carried out to list 31 academic works focused on the use of RVTs to support the teaching of MSE. From this list, an analysis has been made of the areas of MSE on which the works have focused and which RVTs have been used.

2 Methodology

The systematic search [33–35] has been carried out in Scopus, Web of Science and IEEE Xplore, using the following search strings:

- Scopus: TITLE-ABS-KEY(("virtual reality" OR "augmented reality" OR "mixed reality") AND ("materials science" OR "materials engineering" OR "materials science and engineering" OR "materials characterization" OR "materials testing"))
- Web of Science: TS = (("virtual reality" OR "augmented reality" OR "mixed reality") AND ("materials science" OR "materials engineering" OR "materials science and engineering" OR "materials characterization" OR "materials testing"))
- IEEE Xplore: ("All Metadata": "materials science" OR "materials engineering" OR "materials science and engineering" OR "materials characterization" OR "materials testing")

This search was conducted on 24 July 2021 and was restricted to works written in English, published from 2010 onwards in the form of articles, conference papers or book chapters. This generated 269 results, of which 184 were obtained from Scopus, 31 from Web of Science and 54 from IEEE Xplore. After eliminating duplicate results (29 in total), 240 unique results remained, whose titles and abstracts were analysed in detail. This analysis resulted in a total of 31 works describing some form of RVT tools aimed at improving the teaching of MSE. Figure 1 summarises the search, analysis and selection procedure followed during the systematic search that was carried out.

Each of the 31 selected papers was analysed in detail and determined: (i) areas of MSE in which it is circumscribed [36]; and (ii) RVTs employed (VR, AR or MR). Subsequently, this information was analysed in aggregate form in order to determine which areas of MSE have been the most addressed, as well as the RVTs that have been most used. On the basis of this analysis, it has been possible to conclude the current state of the use of RVTs in the teaching of MSE in order to facilitate the opening of new lines of research.

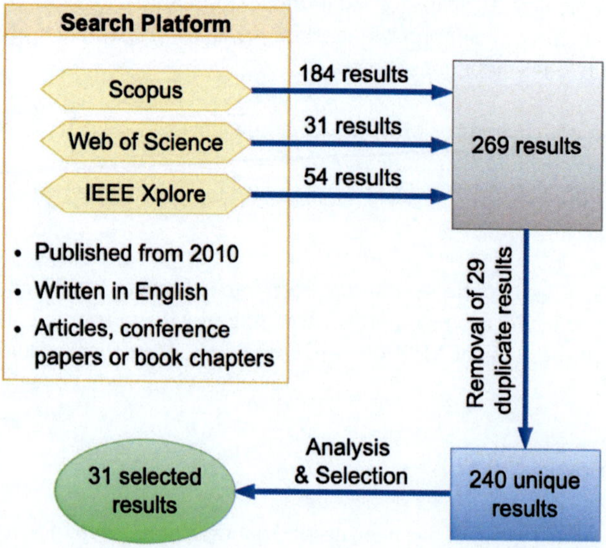

Fig. 1. Flowchart followed to perform the systematic search to select 31 relevant works.

3 Results

3.1 List of Relevant Works and Their Properties

The 31 relevant works selected are listed in Table 1, indicating their year of publication, type of document, country of the authors affiliation and areas of MSE in which they focus and the RVTs employed.

Table 1. List of 31 selected works, year of publication, MSE areas where they are circumscribed and RVTs used.

Work	Year	Document type	Country	MSE area	RVT -used
[37]	2011	Conference paper	USA	Materials structure, processing, and properties	NIVR
[38]	2012	Conference paper	Germany	Biomaterials and soft materials	NIVR
[39]	2013	Article	Portugal	Materials structure, processing, and properties	NIVR
[40]	2013	Conference paper	USA	Materials structure, processing, and properties Material characterization	NIVR
[41]	2014	Article	USA	Materials structure, processing, and properties	NIVR
[42]	2016	Article	Taiwan	Materials structure, processing, and properties Energy and sustainability	NIVR AR
[43]	2017	Article	Spain	Materials structure, processing, and properties	NIVR
[44]	2015	Book chapter	Spain	Materials structure, processing, and properties	NIVR
[45]	2017	Conference paper	Germany	Materials structure, processing, and properties	IVR
[46]	2017	Conference paper	Germany	Materials structure, processing, and properties Materials computing and data science	IVR
[47]	2018	Conference paper	USA	Materials structure, processing, and properties	IVR
[48]	2018	Conference paper	USA	Materials characterization	IVR NIVR
[49]	2018	Conference paper	Germany	Materials structure, processing, and properties	IVR NIVR
[50]	2018	Conference paper	Qatar	Materials structure, processing, and properties	MR
[51]	2019	Conference paper	USA	Carbon-based nanocomposite materials and applications Materials computing and data science	IVR AR

(continued)

Table 1. (*continued*)

Work	Year	Document type	Country	MSE area	RVT -used
[52]	2019	Article	Taiwan	Materials structure, processing, and properties	IVR NIVR
[53]	2019	Conference paper	USA	Materials structure, processing, and properties	IVR
[54]	2019	Article	Brazil	Materials structure, processing, and properties	IVR NIVR
[55]	2019	Conference paper	Spain	Materials structure, processing, and properties	NIVR
[31]	2019	Article	Spain USA	Materials structure, processing, and properties	NIVR
[56]	2019	Article	Germany	Materials structure, processing, and properties Materials computing and data science	IVR
[57]	2019	Conference paper	Spain	Materials structure, processing, and properties	IVR NIVR
[58]	2019	Conference paper	Spain	Materials structure, processing, and properties	NIVR
[59]	2020	Conference paper	Spain USA	Materials structure, processing, and properties	IVR NIVR
[60]	2020	Conference paper	USA	Electronics, optics and quantum	IVR
[61]	2020	Article	Germany New Zealand	Materials structure, processing, and properties	AR
[62]	2020	Conference paper	USA	Materials structure, processing, and properties	NIVR AR
[63]	2020	Article	Spain USA	Materials structure, processing, and properties	NIVR
[64]	2020	Conference paper	UK China	Materials structure, processing, and properties Electronics, optics and quantum	IVR AR
[65]	2020	Article	USA	Materials structure, processing, and properties	IVR AR
[66]	2021	Conference paper	UK	Materials structure, processing, and properties Electronics, optics and quantum	IVR AR

3.2 Quantitative Analysis

As can be seen from the data in Table 1, the areas of MSE that are the focus of the 31 works analysed and the number of times they are addressed (research question 1) are exposed below and shown in Fig. 2:

- Materials structure, processing, and properties: 27 occasions (71.1%).
- Biomaterials and soft materials: 1 occasion (2.6%).
- Materials characterization: 2 occasions (5.3%).
- Energy and sustainability: 1 occasion (2.6%).
- Materials computing and data science: 3 occasions (7.9%).
- Carbon-based nanocomposite materials and application: 1 occasion (2.6%).
- Electronics, optics and quantum: 3 occasions (7.9%).

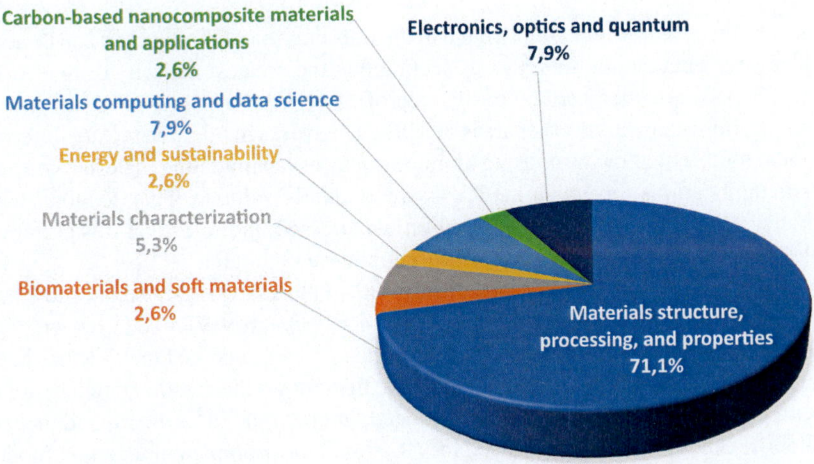

Fig. 2. Distribution of MSE areas addressed in the 31 selected works.

Focusing in the RVTs employed, it can be seen that they have been used (research question 2) as listed below and shown on Fig. 3:

- NIVR: 19 occasions (44.2%)
- IVR: 16 occasions (37.2%)
- AR: 7 occasions (16.3%)
- MR: 1 occasion (2.3%)

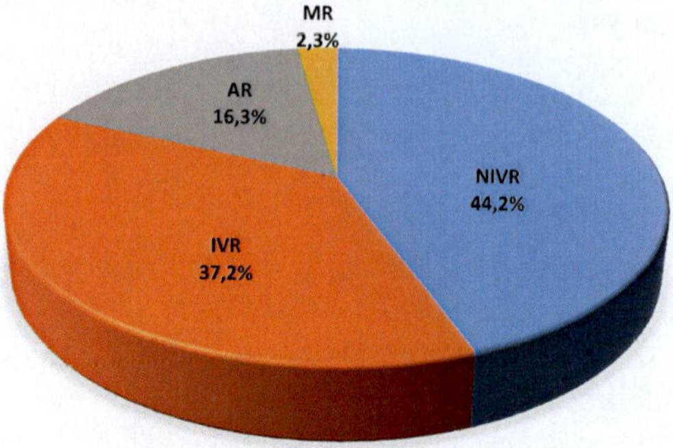

Fig. 3. Distribution of RVTs used in the 31 selected works.

4 Discussion

Regarding to the research question 1 stated in the introduction of this work, it can be seen from the results obtained in the previous section that the greatest majority of the works analysed (71.1%) is circumscribed to the area of *Materials structure, processing, and properties*, followed by far by other areas of MSE. The works in this area mostly describe educational tools that allow visualising atomic structures or simulating materials testing, which are applications for which the RVTs are evidently suitable. This, together with the fact that other areas of MSE have been little addressed, indicates that it is currently possible to open up numerous lines of research in these other areas.

Regarding to the research question 2, the results of the analysis carried out show that the most widely used technology has been VR (81.4% between NIVR and IVR), with AR being far behind (16.3%) and MR being testimonial (2.3%). The fact that VR has been used more than AR may be due to the fact that the first allows the creation of educational tools with greater interaction possibilities. Indeed, interaction via keyboard and mouse, controllers or other control devices used in VR allow more complete interaction modes than those offered by the small touch screens of smartphones and tablets when using AR. Meanwhile, MR is a technology that eliminates the interactivity problems of AR in its current state. However, it is possible to hypothesise that its implementation is still scarce, mainly due to its high price. In this sense, it is to be expected that as MR systems decrease in price, an increasing number of academic works will appear that use this technology to facilitate the teaching of MSE.

5 Conclusions

In this work, a systematic search has been carried out to obtain a picture of the current state of the art in the use of virtual reality technologies applied to the enhancement of Materials Science and Engineering (MSE) teaching. The analysis of 31 papers selected as relevant shows that the main area in which the MSE papers are focused is *Materials structure, processing, and properties*, with the rest of the areas having been little addressed. Therefore, it is possible to explore numerous new lines of research in other areas of MSE. Meanwhile, virtual reality has been the most widely used technology so far, probably due to the greater possibilities of interaction that this technology offers with respect to current augmented reality, which is mainly used in smartphones and tablets. Mixed Reality MR is a currently very little used technology, possibly because it is the technology that has been on the market for the shortest time and is still very expensive.

Acknowledgments. This research has been supported by the project "Intelligent and sustainable mobility supported by multi-agent systems and edge computing (InEDGEMobility): Towards Sustainable Intelligent Mobility: Blockchain-based framework for IoT Security", Reference: RTI2018-095390-B-C32, financed by the Spanish Ministry of Science, Innovation and Universities (MICINN), the State Research Agency (AEI) and the European Regional Development Fund (FEDER).

References

1. Mena Mamani, N.: Machine learning techniques and polygenic risk score application to prediction genetic diseases. Adv. Distrib. Comput. Artif. Intell. J. **9**, 5–14 (2020)
2. Basarslan, M.S., Kayaalp, F.: Sentiment analysis with machine learning methods on social media. Adv. Distrib. Comput. Artif. Intell. J. **9**, 5–15 (2020)
3. Mishra, A., Pathak, T.: Estimation of grain size distribution of friction stir welded joint by using machine learning approach. Adv. Distrib. Comput. Artif. Intell. J. **10**, 99–110 (2020)
4. Karapinar Senturk, Z., Bakay, M.S.: Machine learning based hand gesture recognition via EMG data. Adv. Distrib. Comput. Artif. Intell. J. 10 (2021)
5. Ali, Z., ur Rehman, I., Jaan, Z.: An empirical analysis on software development efforts estimation in machine learning perspective. Adv. Distrib. Comput. Artif. Intell. J. **10**, 227–240 (2021)
6. Mateos García, N.: Multi-agent system for anomaly detection in industry 4.0 using machine learning techniques. Adv. Distrib. Comput. Artif. Intell. J. **8**, 33–40 (2019)
7. Castelltort, A., Chabert, A., Hersog, N., Laurent, A., Sala, M.: Fuzzy rules based solution for system administration security management via a blockchain. Adv. Intell. Syst. Comput. **1010**, 71–78 (2020)
8. Ciatto, G., Maffi, A., Mariani, S., Omicini, A.: Smart contracts are more than objects: pro-activeness on the blockchain. Adv. Intell. Syst. Comput. **1010**, 45–53 (2020)
9. Munoz, D.J., Constantinescu, D.A., Asenjo, R., Fuentes, L.: ClinicAppChain: a low-cost blockchain hyperledger solution for healthcare. Adv. Intell. Syst. Comput. **1010**, 36–44 (2020)
10. Odelu, V.: IMBUA: identity management on blockchain for biometrics-based user authentication. Adv. Intell. Syst. Comput. **1010**, 1–10 (2020)
11. Rosa, J.: On value preservation with distributed ledger technologies, intelligent agents, and digital preservation. Adv. Intell. Syst. Comput. **1010**, 145–152 (2020)
12. Sousa, H., Pinto, A.: Blockchain based informed consent with reputation support. Adv. Intell. Syst. Comput. **1010**, 54–61 (2020)
13. Wunderlich, S., Saive, D.: The electronic bill of lading: challenges of paperless trade. Adv. Intell. Syst. Comput. **1010**, 93–100 (2020)
14. Casado-Vara, R., Novais, P., Gil, A., Prieto, J., Corchado, J.: Distributed continuous-time fault estimation control for multiple devices in IoT networks. IEEE Access **7**, 11972–11984 (2019)
15. Casado-Vara, R., Martin-del Rey, A., Affes, S., Prieto, J., Corchado, J.: IoT network slicing on virtual layers of homogeneous data for improved algorithm operation in smart buildings. Future Gener. Comput. Syst. **102**, 965–977 (2020)

16. Costa, Â., Novais, P., Corchado, J., Neves, J.: Increased performance and better patient attendance in an hospital with the use of smart agendas. Log. J. IGPL **20**, 689–698 (2012)
17. Chamoso, P., González-Briones, A., De La Prieta, F., Venyagamoorthy, G., Corchado, J.: Smart city as a distributed platform: toward a system for citizen-oriented management. Comput. Commun. **152**, 323–332 (2020)
18. González-Briones, A., Chamoso, P., De La Prieta, F., Demazeau, Y., Corchado, J.: Agreement technologies for energy optimization at home. Sensors **18** (2018)
19. Saposnik, G., et al.: Effectiveness of virtual reality using Wii gaming technology in stroke rehabilitation: a pilot randomized clinical trial and proof of principle. Stroke **41**, 1477–1484 (2010)
20. Berg, L.P., Vance, J.M.: Industry use of virtual reality in product design and manufacturing: a survey. Virtual Real. **21**, 1–17 (2017)
21. Muhanna, M.A.: Virtual reality and the CAVE: taxonomy, interaction challenges and research directions. King Saud Univ. Comput. Inf. Sci. **27**, 344–361 (2015)
22. Jensen, L., Konradsen, F.: A review of the use of virtual reality head-mounted displays in education and training. Educ. Inf. Technol. **23**, 1515–1529 (2018)
23. Mosher, M.A., Carreon, A.C.: Teaching social skills to students with autism spectrum disorder through augmented, virtual and mixed reality. Res. Learn. Technol. **29**, 2626 (2021)
24. Carmigniani, J., Furht, B., Anisetti, M., Ceravolo, P., Damiani, E., Ivkovic, M.: Augmented reality technologies, systems and applications. Multimed. Tools. Appl. **51**, 341–377 (2011)
25. Hoyer, W.D., Kroschke, M., Schmitt, B., Kraume, K., Shankar, V.: Transforming the customer experience through new technologies. J. Interact. Mark. **51**, 57–71 (2020)
26. Collins, J., Regenbrecht, H., Langlotz, T.: Visual coherence in mixed reality: a systematic enquiry. Presence Teleoper. Virtual Environ 26, 16–41 (2017)
27. Vergara, D., Antón-Sancho, Á., Extremera, J., Fernández-Arias, P.: Assessment of virtual reality as a didactic resource in higher education. Sustainability **13** (2021)
28. Vergara, D., Rubio, M.P., Lorenzo, M.: On the design of virtual reality learning environments in engineering. Multimodal Technol. Interact. **1**, 11 (2017)
29. Vergara, D., Extremera, J., Rubio, M.P., Dávila, L.P.: The proliferation of virtual laboratories in educational fields. Adv. Distrib. Comput. Artif. Intell. J. **9**, 85–97 (2020)
30. Aljojo, N., et al.: Kids' atlas application to learn about geography and maps. Adv. Distrib. Comput. Artif. Intell. J. **9**(2), 33–48 (2020)
31. Vergara, D., Extremera, J., Rubio, M.P., Dávila, L.P.: Meaningful learning through virtual reality learning environments: a case study in materials engineering. Appl. Sci. **9**, 4625 (2019)
32. Vergara, D., Fernández-Arias, P., Extremera, J., Dávila, L.P., Rubio, M.P.: Educational trends post COVID-19 in engineering: virtual laboratories. Mater. Today Proc. **49**, 155–160 (2022)
33. Ospina-Bohórquez, A., Rodríguez-González, S., Vergara-Rodríguez, D.: On the synergy between virtual reality and multi-agent systems. Sustainability **13** (2021)
34. Kitchenham, B.A., Budgen, D., Pearl Brereton, O.: Using mapping studies as the basis for further research – a participant-observer case study. Inf. Softw. Technol. **53**, 638–651 (2011)
35. Petersen, K., Vakkalanka, S., Kuzniarz, L.: Guidelines for conducting systematic mapping studies in software engineering: an update. Inf. Softw. Technol. **64**, 1–18 (2015)
36. Materials Research Society – 2021 MRS Fall Meeting & Exhibit. mrs.org/fall2021. Accessed 01 July 2021
37. Doblack, B.N., Flores, C., Matlock, T., Dávila, L.P.: The emergence of immersive low-cost 3D virtual reality environments for interactive learning in materials science and engineering. Mater. Res. Soc. Symp. Proc. **1320**, mrsf10-1320-xx04-01 (2011)
38. Martinez, H., von Deimling, C., Ulbrich, H., von Eisenhart-Rothe, R., Burgkart, R.: Real-time 3D visualization in an open architecture of a robotic application in the biomechanics. In: 2012 IEEE International Conference on Robotics and Biomimetics, ROBIO 2012, pp. 1458–1463. IEEE (2012)

39. Restivo, M.T., Lopes, A.M., Padilla, L., Chaves, P., Duarte, T.: Haptic system for determining the young modulus of materials. Int. J. Online Eng. **9**, 68–70 (2013)
40. Ari-Gur, P., et al.: Transforming undergraduate engineering education with 3D virtual reality laboratory. In: 120th ASEE Annual Conference and Exposition. ASEE (2013)
41. Doblack, B.N., Allis, T., Dávila, L.P.: Novel 3D/VR interactive environment for MD simulations, visualization and analysis. J. Vis. Exp. **94**, e51384 (2014)
42. Liou, W.-K., Bhagat, K.K., Chang, C.-Y.: Beyond the flipped classroom: a highly interactive cloud-classroom (HIC) embedded into basic materials science courses. J. Sci. Educ. Technol. **25**(3), 460–473 (2016). https://doi.org/10.1007/s10956-016-9606-8
43. Vergara, D., Rubio, M.P., Lorenzo, M.: New approach for the teaching of concrete compression tests in large groups of engineering students. J. Prof. Issues Eng. Educ. Pract. **143**, 05016009 (2017)
44. Vergara, D., Lorenzo, M., Rubio, M.P.: Virtual environments in materials science and engineering: the students' opinion. In: Handbook of Research on Recent Developments in Materials Science and Corrosion Engineering Education, 1st edn, pp. 148–165. IGI Global, Hershey (2015)
45. Ortelt, T.R., Ruider, E.: Virtual lab for material testing using the oculus rift. In: 4th Experiment@International Conference, exp.at 2017, pp. 145–146. IEEE (2017)
46. García-Hernández, R.J., Kranzlmüller, D.: Virtual reality toolset for material science: NOMAD VR tools. In: De Paolis, L.T., Bourdot, P., Mongelli, A. (eds.) AVR 2017. LNCS, vol. 10324, pp. 309–319. Springer, Cham (2017). https://doi.org/10.1007/978-3-319-60922-5_25
47. Caro, V., Carter, B., Dagli, S., Schissler, M., Millunchick, J.: Can virtual reality enhance learning: a case study in materials science. In: 48th Frontiers in Education Conference, FIE 2018, 8659267. IEEE (2018)
48. Banic, A., Money, J.H., Khadka, R.: Evaluation of scientific workflow effectiveness for a distributed multi-user multi-platform support system for collaborative visualization. In: ACM International Conference Proceeding Series, Proceedings of the 2018 Practice and Experience in Advanced Research Computing Conference: Seamless Creativity, PEARC 2018, 61. ACM (2018)
49. Grodotzki, J., Ortelt, T.R., Tekkaya, A.E.: Remote and virtual labs for engineering education 4.0: achievements of the ELLI project at the TU Dortmund University. In: Wang, L. (ed.) Procedia Manufacturing, Proceedings of the 46th SME North American Manufacturing Research Conference, NAMRC 46, vol. 26, pp. 1349–1360. Elsevier (2018)
50. Mansoor, B., Makki, M.J., Al-Thani, D.: Use of mixed reality tools in introductory materials science courses. In: 125th ASEE Annual Conference and Exposition, 22405. ASEEE (2018)
51. Iqbal, R., Kuttenkuler, T., Brewer, C., Sakidja, R.: A framework for visualizing the dynamic events of carbon nanocomposites using virtual and augmented reality tools. In: Obaidat, M.S., et al. (eds.) 16th International Joint Conference on e-Business and Telecommunications, ICETE 2019, pp. 337–342. SciTePress (2019)
52. Tarng, W., Chen, C.J., Lee, C.Y., Lin, C.M., Lin, Y.J.: Application of virtual reality for learning the material properties of shape memory alloys. Appl. Sci. **9**, 580 (2019)
53. Greenwald, S.W., McDowell, G., Corning, W., Devarakonda, A., Ye, L., Devarakonda, A.: Crystal VR: creating an immersive scientific tool for learning and research. In: 2019 IEEE International Conference on Engineering, Technology and Education, TALE 2019, 9225971. IEEE (2019)
54. Bardella, F., Rodrigues, A.M., Leal, R.M.: CrystalWalk: an educational interactive software for synthesis and visualization of crystal structures. J. Mater. Educ. **41**, 157–180 (2019)
55. Extremera, J., Vergara, D., Rubio, M.P., Gómez, A.I.: Design of virtual reality learning environments: step-by-step guidance. In: 12th Annual International Conference of Education, Research and Innovation, ICERI 2019, pp. 1285–1290. IATED (2019)

56. García-Hernández, R.J., Kranzlmüller, D.: NOMAD VR: multiplatform virtual reality viewer for chemistry simulations. Comput. Phys. Commun. **237**, 230–237 (2019)
57. Vergara, D., Sánchez, M., Garcinuño, A., Rubio, M.P., Extremera, J., Gómez, A.I.: Spatial comprehension of crystal lattices through virtual reality applications. In: 12th Annual International Conference of Education, Research and Innovation, ICERI 2019, pp. 1291–1295. IATED (2019)
58. Rubio, M.P., Vergara, D., Rodríguez, S., Extremera, J.: Virtual reality learning environments in materials engineering: rockwell hardness test. In: Di Mascio, T., et al. (eds.) MIS4TEL 2018. AISC, vol. 804, pp. 106–113. Springer, Cham (2019). https://doi.org/10.1007/978-3-319-98872-6_13
59. Extremera, J., Vergara, D., Rubio, M.P., Dávila, L.P., De la Prieta, F.: Effects of time in virtual reality learning environments linked with materials science and engineering. In: Vittorini, P., Di Mascio, T., Tarantino, L., Temperini, M., Gennari, R., De la Prieta, F. (eds.) MIS4TEL 2020. AISC, vol. 1241, pp. 1–9. Springer, Cham (2020). https://doi.org/10.1007/978-3-030-52538-5_1
60. Batra, J.S., Richardson, R., Webb, R.: How can instructors strengthen students' motivation to learn complex 3D concepts in an engineering classroom? In: 2020 IEEE Frontiers in Education Conference, FIE 2020, 9274193. IEEE (2020)
61. Müssig, J., Clark, A., Hoermann, S., Loporcaro, G., Loporcaro, C., Huber, T.: Imparting materials science knowledge in the field of the crystal structure of metals in times of online teaching: a novel online laboratory teaching concept with an augmented reality application. J. Chem. Educ. **97**, 2643–2650 (2020)
62. Hain, A., Motaref, S.: Implementing interactive 3-D models in an entry level engineering course to enhance students' visualization. In: 2020 ASEE Virtual Annual Conference, ASEE 2020, 825. ASEE (2020)
63. Vergara, D., Extremera, J., Rubio, M.P., Dávila, L.P.: The technological obsolescence of virtual reality learning environments. Appl. Sci. **10**, 915 (2020)
64. Bourguet, M.L., Wang, X., Ran, Y., Zhou, Z., Zhang, Y., Romero-Gonzalez, M.: Virtual and augmented reality for teaching materials science: a students as partners and as producers project. In: Mitsuhara, H., et al. (eds.) 2020 IEEE International Conference on Teaching, Assessment, and Learning for Engineering, TALE 2020, pp. 452–459. IEEE (2020)
65. Srinivasa, A.R., Jha, R., Ozkan, T., Wang, Z.: Virtual reality and its role in improving student knowledge, self-efficacy, and attitude in the materials testing laboratory. Int. J. Mech. Eng. Educ. **49**, 382–409 (2020)
66. Bourguet, M.L., Romero-Gonzalez, M.: Work-in-progress—teaching invisible phenomena and virtual experiments: immersion or augmentation? In: Economou, D., et al. (eds.) 7th International Conference of the Immersive Learning Research Network, iLRN 2021, 9459308. IEEE (2021)

Kaleidoscope: A Multi-perspective Technology-Enhanced Observation Method to Support the Development of Negotiation Skills

Leonardo Caporarello[1](✉) and Stefano Magoni[2]

[1] SDA Bocconi School of Management, and Bocconi University, Department of Management and Technology, and Bocconi University Innovations in Learning and Teaching (BUILT), Milan, Italy
leonardo.caporarello@unibocconi.it

[2] Psychologist, Bocconi University Innovations in Learning and Teaching (BUILT), Milan, Italy
stefano.magoni@unibocconi.it

Abstract. Covid 19 Pandemic forced professors and students to deal with new ways of learning, often enhanced by technological solutions. Other than a mere response to the emergency, this pushes to new digital solution turned out to be a big opportunity to discover new learning methods. In this paper we present Kaleidoscope, an innovative educational method based on a multi-dimensional observation of a negotiation simulation. Thanks to the technological solutions, Kaleidoscope gather data on participants' experience from different sources with the aim of unveiling cognitive, emotional, and relational dynamics that occurs during a negotiation process. The learning purpose is to increase the learners' awareness on these dimensions by using a live debrief dashboard that shows the relationship between the different variables detected. Today Kaleidoscope has yet to go live in a real classroom, it has only been partially tested in a laboratory context. In this paper we present the learning objectives of this project, the theoretical framework with the dimension considered. Finally, we acknowledge the current limitations of this innovative method and the future steps.

Keywords: Educational method · Experiential learning · Technology · Emotions · Facial expression · Negotiation · Simulation

1 Introduction

The Covid 19 Pandemic forced educational institutions to rethink the learning experience through new pedagogical methods, often requiring a technological support. Although the "Time for Higher Education to Prioritize Digital Strategy" [1] has arrived, innovation cannot emerge from the mere use of new technological tools. As well explained by the TPACK framework [2], there must be an agreement between the pedagogical, technological, and learning dimensions.

In this context, we intend to investigate how to use digital technology to create a new educational experience that could effectively empower learners to develop their

M. Temperini et al. (Eds.): MIS4TEL 2022, LNNS 580, pp. 59–64, 2023.
https://doi.org/10.1007/978-3-031-20617-7_8

behavioral skills. We designed a method that can be applied in the online setting as well to the context that now represents the "new normality": the "phygital" one.

We put together a team that managed all the aspects of this project by investigating the design elements: from a methodological, technological, legal, and, of course, scientific-content point of view. The result was the design of Kaleidoscope: an innovative educational method based on a multi-dimensional observation (self, others, tech-based) of social interaction dynamics within "Negotiation skills" courses (at undergraduate, graduate and executive levels), specifically the negotiation live meeting. Kaleidoscope allows to collect the observations from different points of view, and through a live dashboard elaborates and highlights some analyses in terms of behaviors and dynamics that emerge during the meeting, based on the dimensions of the underlying theoretical model.

The data are collected through different methods, including questionnaires, direct observations (self-, peers and by the other party), and recognition of the emotional states through a facial expression recognition application [3, 4]. The technological component, therefore, becomes essential at least to:

- Allow faculty members to create an interactive, engaging, and enjoyable educational experience.
- Guide learners throughout the learning experience.
- Generate real-time graphs and evidence useful both to provide students with continuous feedback (a personalized report is generated), and the debriefing session.

The simulation of a negotiation process, considered as part of the broader frame of experiential learning [5], aims to stimulate learners to apply the knowledge and skills acquired during their educational path, through an experience of Team Based-Active Learning.

The next section presents the learning objectives of Kaleidoscope. The third section describes the theoretical reference model and the variables taken into consideration to observe the phenomenon. Finally, limitations and recommendations for future research are presented.

2 Kaleidoscope: Intended Learning Objectives

Three are the main Kaleidoscope's intended learning objectives. We describe them in this section.

2.1 Objective 1: Shed Light on the Complex Dynamics of a Negotiation Process

The human being can only be grasped in his emotional and relational nature, regardless of the context of observation. Each negotiation process must be considered within a relational frame that involves at least two parties involved. These premises justify the huge number of studies in the literature that describe the role of numerous variables, that can influence the choices made by negotiators on their way to an agreement. Among these, there are emotions [6] and self-efficacy [7, 8]. Reference literature has demonstrated that these factors can play an important part in negotiation dynamics [9].

Kaleidoscope focuses precisely on these behavioral variables, showing how negotiators' emotions, beliefs, and assumptions can impact the objective and subjective outcomes of the negotiation process.

2.2 Objective 2: Increasing the Level of Preparation on Such Dynamics

To make an effective communication when negotiating, parties must first establish a relationship between them. To that end, understanding each party's negotiation style can be very useful. Only by acknowledging their emotional attitudes and negotiating style(s), negotiators will be able to better explore the situation (mainly, context, mandate, profiles of all the parties involved in the process) and being eventually more prepared to achieve higher performance [10]. Kaleidoscope allows learners to deepen the knowledge of their own, as other party's as well, negotiating style(s), taking into consideration emotional and relational dynamics. Schlegel et al. [11] have investigated how emotional intelligence and some emotions recognition abilities could be predictive of negotiation outcomes. Thanks to Kaleidoscope the learner will have the opportunity to get to know themselves better not only as a negotiator (as professional) but also as individuals (as persons). With this purpose, Kaleidoscope requires learners to observe the phenomenon of the negotiation meeting through different observation levels, e.g. 1. self-report, 2. teammates' perception, 3. other spokesperson's perception, 4. other spokesperson's teammates perception, 5. technology-based, 6. human impartial observer.

2.3 Objective 3: Enhancing Learners' Engagement

Kaleidoscope is a team-based educational method. Although the negotiation meeting will be carried out by a single spokesperson, who is nominated within the team, the success or failure of the negotiation should be considered as a result of a team effort. Indeed, other than the spokesperson, every team member can play a specific role, which may include e.g. time keeper, observers, and data analyst. Consistently with that objective, Kaleidoscope allows also to monitor and collect continuous feedback (at some specific time frame) about the process and outcome satisfaction [12] so to highlight the contribution and effort by the team as a whole.

3 Theoretical Framework

This section presents the theoretical model and the instruments for each variable.

3.1 Negotiation Strategy and Style

Literature distinguishes the negotiators' orientation, defined as a propensity towards specific objectives and behaviors, into "distributive" and "integrative" [13]. Several studies have brought empirical evidence on how the chosen strategy and its tactics can influence the negotiation's outcome [14]. Craver states that the most effective negotiators are the ones able to employ a hybrid competitive (distributive) problem-solving (integrative) approach, thus managing to adapt to the context [10]. To identify the learners' styles, we use the MODE of Kilmann and Thomas [15]. Furthermore, just before starting the negotiation meeting, participants are asked to specify the strategy they intend to adopt, and the strategy they have effectively applied.

3.2 Self-efficacy

The concept of self-efficacy, developed by Bandura, refers to the belief in one's ability to organize and carry out the course of action necessary to adequately manage the situations encountered and to achieve the desired results. These beliefs of efficacy influence how people think, feel, find personal motivations, and act [7, 8]. Several studies found that negotiator's self-efficacy can influence negotiation final outcomes [9, 16].

This variable is measured using the New General Self-Efficacy Scale (NGSE) [17].

3.3 Affect and Relational Dimensions

In this study emotions are measured through the "Affect" construct, as we considered it more informative for learners than, for example, basic emotions. We use two different instruments to measure this construct. The first instrument is the Positive and Negative Affective Schedule – PANAS – [18], which investigates two factors: the Positive Affect measures the engagement and positive feelings at the time of the measurement, while the Negative Affect measures negative feelings and distress.

As an "innovation within the innovation" of this educational method, we have also decided to measure emotions through the recognition of facial expressions, using a software able to detect the Arousal and Pleasantness value associated with the face expressions. Evidence from this type of observation can be analyzed according to Circumplex Model of the Affect [19], thus allowing comparisons with PANAS results.

In addition to the possibility of identifying certain "emotional action-reaction" events, we investigate two variables created ad-hoc for this methodology and measured through the PANAS deltas. The first will be called "Emotional Synchrony", which evaluates how similar the experiences of the spokespersons are, derived from the deltas of the PANAS self-referred values. The second variable is "Perception/Empathic Accuracy", defined by the mean of the PANAS deltas between what the spokesperson experiences and the other parties' perception of their emotional state.

3.4 Team Effectiveness

Mehta et al. [20] state that this variable goes beyond the quantitative performance, and includes qualitative dimensions as well: team viability, team satisfaction, and team perceived performance defined as a sense of how well the team is doing [21]. In addition to showing the group's judgment on internal team dynamics, a greater Team Effectiveness can also lead to better learning because group members are more likely to assist and cooperate through communication of ideas and the exchange of knowledge [22].

To measure this variable, we have selected and adapted some items from the Teamwork Quality (TQM) framework proposed by Hoegl and Gemuenden [23].

3.5 Satisfaction

Although the level of satisfaction is strongly correlated to the results of the negotiation, several studies have shown how this dimension presents different nuances and elements: for example, Pruitt [24] suggests that satisfaction can be influenced not only by the personal gains but also from the "Joint gains" of the parties and, furthermore, by the perception of the differences between the own results and those of the others.

We measure participants' satisfaction by using the Subjective Value Inventory (SVI) of Curhan et al. [12].

4 Final Remarks, Limitations, and Future Steps

The educational method described in this paper has already been tested inside a laboratory context. From the research point of view, we can define the "Kaleidoscope method" as a quasi-experiment.

There are few technological aspects on which a fine tuning is necessary, but we are confident that this experience will go live for the first time on winter 2022. We took in consideration different expression recognition software such as FaceReader (Noldus), Morphcast, OpenFace 2.2.0 and other open-source software. The software we intend to use to manage the groups will be produced ad hoc with our technological partner. The next steps for this project will consist in adopting in a quite extensive way this method inside real classrooms (as for today, it has been tested in one "pilot" class), which may result in additional corrections to better shape the experience, and publishing the results gathered to contribute to the research on this matter.

References

1. Gallagher, S., Palmer, J.: The Pandemic Pushed Universities Online. The Change Was Long Overdue (2020). https://hbr.org/2020/09/the-pandemic-pushed-universities-online-the-change-was-long-overdue
2. Koehler, M.J., Mishra, P.: What happens when teachers design educational technology? The development of technological pedagogical content knowledge. J. Educ. Comput. Res. **32**(2), 131–152 (2005)
3. Ekman, P.: Universals and cultural differences in facial expressions of emotions. In: Cole, J. (ed.) Nebraska Symposium on Motivation, pp. 207–282. University of Nebraska Press, Lincoln (1972)
4. Dupré, D., Krumhuber, E.G., Küster, D., McKeown, G.: A performance comparison of eight commercially available automatic classifiers for facial affect recognition. PLOS One **15**(4), e0231968 (2020)
5. Kolb, D.A.: Experiential Learning: Experience as the Source of Learning and Development. Prentice-Hall, Inc., Englewood Cliffs (1984)
6. Olekalns, M., Druckman, D.: With feeling: how emotions shape negotiation. Negot. J. **30**, 455–478 (2014)
7. Bandura, A.: Self-efficacy: toward a unifying theory of behavioral change. Psychol. Rev. **84**, 191–215 (1977)
8. Bandura, A.: Self-efficacy mechanism in human agency. Am. Psychol. **37**, 122–147 (1982)

9. Brett, J.F., Pinkley, R.L., Jackofsky, E.F.: Alternatives to having a BATNA in dyadic negotiation: the influence of goals, self-efficacy, and alternatives on negotiated outcomes. Int. J. Confl. Manag. **7**, 121–138 (1996). https://doi.org/10.1108/eb022778
10. Craver, C.: Negotiation styles: the impact of bargaining transactions. Dispute Resolut. J. **58**(1), 48–56 (2003)
11. Schlegel, K., Vicaria, I.M., Isaacowitz, D.M., Hall, J.A.: Effectiveness of a short audiovisual emotion recognition training program in adults. Motiv. Emot. **41**(5), 646–660 (2017). https://doi.org/10.1007/s11031-017-9631-9
12. Curhan, J.R., Elfenbein, H.A., Xu, H.: What do people value when they negotiate? Mapping the domain of subjective value in negotiation. J. Personal Soc. Psychol **91**(3), 493–512 (2006)
13. Walton, R.E., McKersie, R.B.: A Behavioral Theory of Labor Negotiations. McGraw-Hill, New York (1965)
14. Kern, M., Brett, J., Weingart, L.: Getting the floor: motive-consistent strategy and individual outcomes in multi-party negotiations. Group Decis. Negot. **14**(1), 21–41 (2005). https://doi.org/10.1007/s10726-005-3874-7
15. Thomas, K.W., Kilmann, R.H.: Developing a forced-choice measure of conflict behavior: the "MODE" instrument. Educ. Psychol. Measur. **37**, 309–325 (1977)
16. Miles, E.W., Maurer, T.J.: Advancing validity of self-efficacy in negotiation through focusing at the domain level. J. Occup. Organ. Psychol. **85**(1), 23–41 (2012)
17. Chen, G., Gully, S.M., Eden, D.: Validation of a new general self-efficacy scale. Organ. Res. Methods **41**, 62–83 (2001)
18. Watson, D., Clark, L.A., Tellegen, A.: Development and validation of brief measures of positive and negative affect: the PANAS scales. J. Pers. Soc. Psychol. **54**, 1063–1070 (1988)
19. Russell, J.A.: A circumplex model of affect. J. Pers. Soc. Psychol. **39**, 1161–1178 (1980)
20. Mehta, A., Mehta, N.: Knowledge integration and team effectiveness: a team goal orientation approach. Decis. Sci. **49**(3), 445–486 (2018)
21. Hackman, J.R.: Groups That Work (and Those That Don't). Jossey Bass, San Francisco (1990)
22. Lin, C.Y., Huang, C.K.: Understanding the antecedents of knowledge sharing behaviour and its relationship to team effectiveness and individual learning. Australas. J. Educ. Technol. **36**(2), 89–104 (2020)
23. Hoegl, M., Gemuenden, H.G.: Teamwork quality and the success of innovative projects: a theoretical concept and empirical evidence. Organ. Sci. **12**, 435–449 (2001)
24. Pruitt, D.G.: Negotiation Behavior. Academic Press Inc, New York (1981)

A New Metric to Help Teachers Unveil Meaningful Learning in Concept Maps

Antonio Fabrizio Fiume[1], Filippo Sciarrone[1](✉), and Marco Temperini[2]

[1] Faculty of Economics, Universitas Mercatorum, Piazza Mattei, 10, Rome, Italy
antoniofabriziofiume@libero.it, filippo.sciarrone@unimercatorum.it
[2] Department of Computer, Control and Management Engineering, Sapienza University of Rome, Via Ariosto, 25, Rome, Italy
marte@diag.uniroma1.it

Abstract. Concept Maps are effective tools to represent and share one's vision about a particular knowledge domain or topic, in a personal, graphic and structured way. They are widely used and recommended in the educational field to both students and teachers. Most of students use Concept Maps at school, to better represent and share their knowledge, and to represent learning about a particular topic. Moreover, teachers take advantage of the potential of Concept Maps, to evaluate students, also with respect to the occurrence of meaningful learning. Here we propose a new metrics (the *Concept Entropy*), devised to provide a teacher with significant information about the topological placement of concepts in a Concept Map, and the possible occurrence of meaningful learning. An experimental activity on a selection of basic Concept Maps structures shows the usefulness of the new metrics, by comparing it with two baseline metrics.

Keywords: Concept Maps · Meaningful learning · Concept maps metrics

1 Motivations and Goals

Living in the social-digital age means to be continuously exposed to stimuli arriving from many internet based sources. People process a lot of information every day, so that putting facts and ideas in order has become a common need. At school, university or at work, Concept Maps (CMs) [19,20] are very useful tools, helping to represent and deal with information, and allowing the development and sharing of one's vision about a knowledge domain. For this reason, CMs are widely used and recommended in the educational field both for students and teachers: many students use CMs at school to represent and share their learning process; teachers can take advantage of the potentials of concept mapping (i.e., of the activity of drawing and updating a CM) to organize their curricula and courses [2,9,13], to populate their courses with suitable learning material [8,16], and eventually to evaluate students' skills [1,7,24]. The

M. Temperini et al. (Eds.): MIS4TEL 2022, LNNS 580, pp. 65–74, 2023.
https://doi.org/10.1007/978-3-031-20617-7_9

definition of CMs as effective representations of knowledge domains is in the information processing theory of learning [21]. By a CM knowledge is represented as a network of connected concepts, and when an individual (such as a student) develops such a network, it is likely to be unique, as unique is the personal experience of such individual. However such a network is not static: it is updated whenever new knowledge is learned. This use of CMs in education is strongly related with the concepts of *meaningful* and *rote* learning, introduced in the learning theory of David Ausubel. Such theory is actually based on the idea that knowledge (what has been already learned) is organized in a hierarchy, and that meaningful learning of new knowledge is a process integrating the new into the already organized knowledge, extending the hierarchy. Meaningful learning is based on an active, engaged, and constructive approach, by the part of the student, and is long-lasting. On the contrary, *rote learning*, i.e. the accumulation of information by means of memorization, without processing it and relating it to prior knowledge [4]. From a structural point of view, a CM is represented by a Direct Acyclic Graph (DAG) where nodes are concepts and edges represent semantic relationships. Moreover, graph-level statistics can capture the relationship between graph-level structure and information flow in graphs. In fact, high interconnectivity in a graph corresponds to shared knowledge, better information transmission and the rate of information spread [10,11]. Under the assumption that the learning process of a student is well rendered by the CM that represents the student's knowledge domain, in this paper we propose a new metrics, called *Concepts Entropy* CH, whose name is inspired by the notion of *Entropy* - H - as introduced in Shannon's *Information Theory* [22]), assessing how meaningful is the student's learning according to the related CM; We present a first study of CH, based on meaningful case studies. The paper is structured as follows. Section 2 shows some important background and related works. Section 3 introduces the CH metrics. In Sect. 4 the system used to test the new metrics is shown, while in Sect. 5 are evaluated some significant case studies. Finally the last section discusses future work and conclusions.

2 Background and Related Work

This section provides the reader with some background about 1) the pedagogical implications of CMs' usage, and 2) some of the metrics proposed in literature to assess CMs characteristics, comprising a comparison with the CH proposed in this paper.

2.1 Concept Maps from a Pedagogical Point of View

CMs are hierarchical tools for organizing and representing knowledge in a simple, personal and graphic way, proposed in the late 1960s s and early 1970s from Joseph D. Novak [19–21]. They allow to point out the concepts of (a part of

a) Knowledge Domain, together with semantic connections holding among such concepts. The connections between concepts allow to study how the information related to the knowledge domain not only is represented by the CM, but is also flowing through the CM, in turn representing how the knowledge domain is interpreted by the CM designer: a CM represents knowledge, and also lets the reader reconstruct the reasoning, about the knowledge domain, of the CM's author. Consequently, the typical structure of a CM makes it a useful and valid tool to support meaningful learning: for a learner, it can give a guide along the study, when it is already drawn, and can support the study and the research of information, when it is built by the student her/himself; for the teacher it is a visual aid to explain and analyze the knowledge domain [18], and is also a means to learn about the student and the functioning of her/his mind, as well as assess the learner's cognition (understanding, and mastering, of the knowledge domain), and meta-cognitive abilities [3,5].

2.2 Concept Maps from a Structural Point of View

Several metrics have been developed to allow assessing CMs. They are mainly based on the DAG structure of a CM, with nodes representing concepts, and edges representing semantic relationships between concepts. Traditional approaches rely on counting various elements or features of a CM (e.g., number of concepts, links, or hierarchies) or the computation of map descriptors (e.g., map density, map complexity, link similarity, overlapping of maps) as functions of these elements [12]. *Holistic* approaches intend to measure the accuracy, width, and depth of the students' understanding, based on an analysis of CMs defined by the students [6]. In [16], a set of pedagogical-based similarity measures is proposed, with the aim to allow a teacher to compare an own CM with other CMs retrieved from the Internet, and help the process of finalization of the teacher's CM, and possibly its implementation through suitable and verified learning materials already available. The proposed similarity criteria and measures, do not take into account the support to meaningful learning. A comparison of CMs is involved in [15], which proposes a way of suggesting additional concepts and learning materials to the teacher during the creation of a CM. This is based on detecting and evaluating concept similarity between the teacher's CM and other CMs that are used for comparison. Topological similarity between CMs is computed by a usefulness measure, which evaluates the structural (i.e., contextual) correspondence between the CMs, and assesses the viability of the connections between the concepts (i.e., between the related learning materials) in the maps. In [17] the authors present a method for matching elements or parts of CMs, based on a similarity flooding algorithm with the aim of supporting the comparison and merging of maps. Measures of CM properties, related to the support of meaningful learning, are less frequently discussed in literature, so it appeared useful to investigate about the usage of a dedicated metrics, i.e. the CH presented in this paper. While founded on the Ausubel's theory of learning, and so related to the analysis of meaningful learning, our *Concept Entropy* metric is also useful to both evaluate the features of a single CM, and the similarity of a CM with another, based on the information spread among concepts and the CM's hierarchical structure.

3 The *Concept Entropy* Measure

Meaningful learning can be seen as an activity in which a CM, representative of the knowledge possessed by a student about a knowledge domain, is transformed in a new, more complex, CM, representative of the addition of concepts and new connections between concepts that the students was able to perform [5]. The definition of the *Concept Entropy* metrics started from the twofold idea of measuring 1) how the different placement of the same concepts, in two different maps, can represent different learning experiences, just for how the paths of concepts are traversed during the learning, and 2) evaluating the occurrence of meaningful learning in a student's CM, based on the presence of cross-links between concepts belonging to separate paths of concepts in the map. Given the two CMs shown in Fig. 1, with the same number of links and concepts, the two standard (counting) metrics that take these characteristics into account are: *Density* D and *Complexity* Δ, defined respectively as follows: $\Delta = \frac{L}{C}$ and $C = \frac{L}{L(L-1)}$, being L the total number of links and C the total number of concepts. These metrics do not change for the two CMs. In particular, Fig. 1(a) shows a CM in a *chain* description of a knowledge domain [14]. A chain CM reflects an attempt to organize the knowledge, but it does not capture the complexity and relations between concepts. Moreover, this knowledge representation cannot cope with new additions and it would be difficult to reorganize in light of new information [23]. Figure 1(b) represents a CM with the same concepts, but arranged in a *spoke* configuration, that is, where a unique *hub* concept exists (see Sect. 5 for a slightly deeper classification of CMs according to [14]).

Definition 1. Given a CM, We define the Concept Entropy $CH(c)$ associated to a generic node c_i as follows: $CH(c_i) = \sum_{j=1}^{n} \frac{1}{|Edge(c_j)_{out}|} \log_2 \frac{1}{|Edge_{out(c_j)}|}$

where $|Edge(c_i)|$ is the number of outgoing links connecting c_i to other concepts.

Definition 2. Given a CM, We define the *Concept Entropy* $CH(CM)$ associated to the entire CM as follows: $CH(CM) = \sum_{i=1}^{n} CH(c_i)$

Notice that, if a CM has n concepts, then the maximum value $CH_{max}(CM)$, for each concept c is obtained when each concept c has the maximum number of outgoing links. Considering the DAG topological aspect, we have: with $CH_{max}(CM) = \log_2[(n-1)!]$. Moreover, if a concept c has no outgoing links, we assume: $CH(c) = 0$). Under the CH measure we have that $CH = 0$ for the *chain* configuration, while $CH = 2$ for the *spoke* configuration, apparently capturing a difference in terms of complexity of the learning experience. In Sect. 5 we will see how the sensibility of the CH metrics is greater than that of baseline metrics, and how we can extend the use of this metrics to an initial approach to the analysis of meaningful learning.

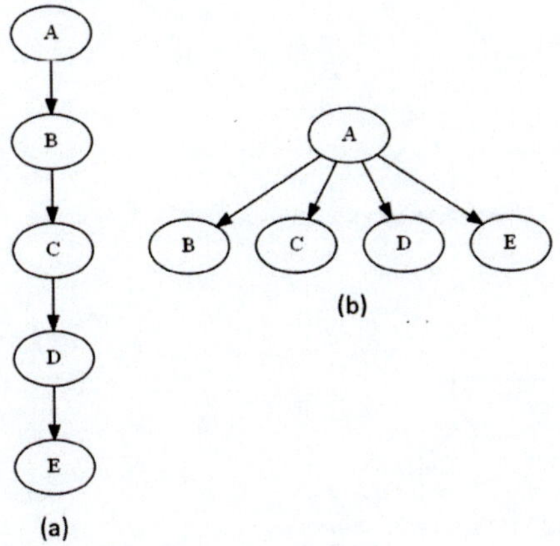

Fig. 1. In figure (a) the CM with chain configuration. In figure (b) a CM with the same concepts but spoke configuration.

4 The *HCM* System

In this section, we describe a web system, called *HCM*, capable to build new CMs together with their *CH* values[1]. By the *HCM* system, the user can perform the following activities:

- *Build a new CM.* Here the user can build a new CM by performing the following steps:
  1. *Add Node.* Here the user can add a new concept.
  2. *Add Edge.* The user can connect two concepts, by selecting the starting node and the ending node.
  3. *Edit Nodes/Edges.* The user can edit a label for each concept, and associate a preposition to each edge. Consequently, the system computes the *CH* measure for each edge;
- *Download the map.* Here the user can download the CM in both JSON and png format;
- *Upload a map.* Here the user can upload a CM, in JSON format, previously downloaded.

Figure 2 shows an example a CM built through the *HCM* system. After entering the concepts and edges, the system will show the new CM, where the concepts are represented by circles and the links between concepts by arrows. The *CH* value of each link is showed along the arrow.

5 Case Studies

In this section we show some comparisons between the *CH* metrics and baseline metrics defined in literature (in particular the *Map Density* and the *Map Complexity* [12]). We try to analyse maps' characteristics from both the structural

[1] Available at: https://entropy-henna.vercel.app/.

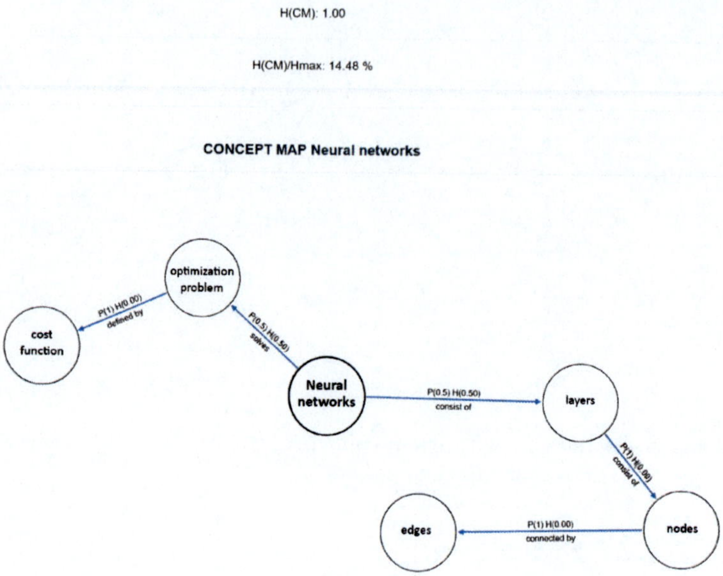

Fig. 2. An example of a CM, built through the HCM system: the neural network domain. The *CH* values are computed and written along each link.

and pedagogic point of view. As shown in Sect. 3, the *CH* metrics gives indications about both the knowledge paths existing through the concepts of a map, and the topological configuration of the CM.

5.1 Case Study 1: Hierarchical Concept Maps

A CM presents a hierarchical structure when it is represented by a tree, with: $L = C - 1$, where L is the number of edges (links from a concept to another) and C is the number of concepts in the CM. Basically we can think of a hierarchical CM as one where each concept is reachable through a unique path of concepts, starting from a root node (concept). Figure 3 shows four typical [14] hierarchical CM architectures: in figure, the part (a) is a *chain* CM; (b), is a *spoke* CM, that is a CM having one concept connected to all the other ones; furthermore, considering that the *level* of a concept in a CM (and of a node in a tree) is the number of edges along the path from the root to the concept, (c) is a (typical [14]) *hierarchical* CM where the concepts have level ≤ 2; finally, in Fig. 3, the CM (d) is still a *hierarchical* CM with concepts having level ≤ 2, yet showing different connections between concepts. Table 1 allows to compare the values of the baseline metrics with those of CH, as they are applied to the four hierarchical maps. While the CH metrics does change, according to different concepts arrangements, the other two metrics offer constant values. We think that the above case study, although limited, can allow to expect CH as a

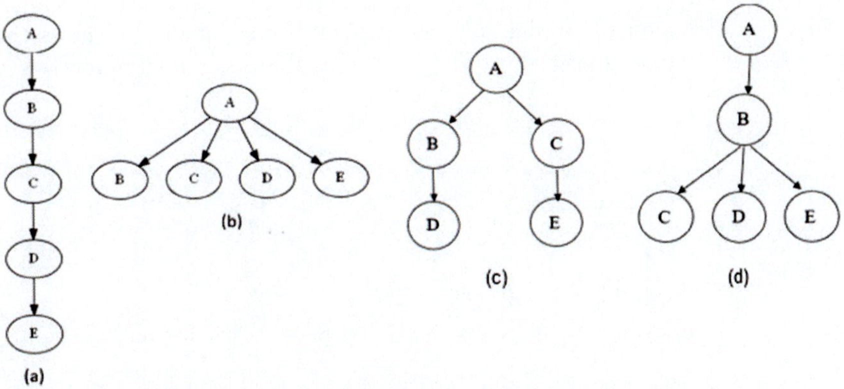

Fig. 3. The sample configurations.

metrics useful for the analysis of hierarchical CMs, as it appears to be sensitive to different arrangements of the same concepts.

Table 1. A comparison among the metrics, on maps having the same concepts and the same number of links. Only the CH metrics is sensitive enough to show different values for the four configurations.

CM	Map density	Map complexity	Concept entropy
(a)	0.2	0.8	0
(b)	0.2	0.8	2
(c)	0.2	0.8	1
(d)	0.2	0.8	1.58

5.2 Case 2: Networked Concept Maps

The second case study shows the usefulness of the CH metrics in more complex CMs, characterized from containing cross-links among concepts belonging to different paths of concepts (paths that could be interpreted as knowledge sub-domains of the whole map). In this case, the CMs shows a more complex and complete knowledge representation than the maps in the previous case study: the cross-links can be interpreted as the occurrence of meaningful learning, i.e. as the addition of new knowledge on a previously current framework of knowledge for the student [3]. In Fig. 4 we show two CMs, made of the same concepts of the ones used in the previous case study: The first CM, labelled as (a), presents a cross-link between a leaf concept and another concept belonging to a different path (still stemming from concept A). The second CM, labelled as (b), presents a cross-link between a concept which is a leaf concept and another concept of different level in the topology of the CM. In Table 2 we give the values of the baseline and CH metrics, applied on the maps of Fig. 4. The table shows that CH reveals to be more sensitive to the differences than the baseline metrics.

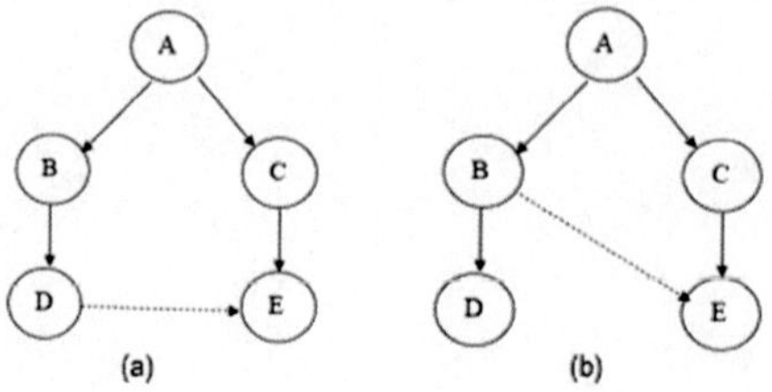

Fig. 4. Two examples of CMs having cross-links.

Table 2. The values of the CH metrics for the networked configurations Vs. the baseline metrics for the same configurations.

CM	Map density	Map complexity	Concept entropy
(a)	0.5	1.0	1
(b)	0.5	1.0	2

5.3 Discussion

The two case studies represent some standard situations that can be found in CMs, and that we used to compare the CH metrics Vs. two baseline metrics: *Complexity* and *Density*. In the first case study, we have proposed four hierarchical CMs, showing that the CH metrics changes in all of them, while the two baseline metrics do not change. In such standard topologies, therefore, the CH metrics is able to capture significant information on the differences among the four configurations used in the case study. From a pedagogical point of view, this means that the CH metrics highlights some differences on the paths that the maps allow to use among the concepts. In fact, if a concept has only one outgoing link, this link will weigh zero, for CH, as it is linked only to another concept, showing a constraint path. In the second case study, the sample CMs present cross-links, i.e., links connecting concepts belonging to different paths of concepts, producing a non-hierarchical structure. As we have seen, the CH metrics reacts to the presence of cross-links, i.e. the additions to already possessed knowledge which enrich the overall knowledge represented in the CM. The experience we are reporting is admittedly limited, and can only be indicative of a significance of the new metrics in so far it shows to be more sensitive than the baseline metrics to typical CM architectures. Moreover we described a metrics that, on the one hand can be used directly on a map to assess some of its characteristics (namely the variety of the paths that the learning experience can follow while traversing the map), while, on the other hand, it can be expected to help ascertain the occurrence of meaningful learning (only as occurrence of new cross-links) when the sequence of CM produced by the students can be managed, that is in a more complex functional architecture than the one of the system currently available (described in Sect. 4. The study we have presented is in its first stage of development and investigation. In fact, the two case studies provide only initial indications on the goodness of the CH metrics with respect to its connection with meaningful learning. The metrics has been shown to be sensitive to structural, i.e. topological, differences presented by a small sample of standard CM configurations. In light of these first positive indications, we think it is worthwhile to continue studying this metrics, using the CHM platform, created specifically for its study, together with a sample of both students and teachers to verify the correlation between it and the meaningful learning.

6 Conclusions and Future Work

We have presented the initial development of a new metric (*Concept Entropy*), designed to measure the propagation of knowledge through the concepts of a CM, in order to unveil the occurrence of meaningful learning. The metrics value, for a CM, depends on the number of connections of concepts with other concepts, and capture the presence of cross-links between concepts listed in different paths of the map. Two case studies were proposed, in order to compare the new metrics with two baseline ones, on basic CM topologies. As a future development we plan both a more in-depth study with other more complex CMs, and a field experimentation implying the collaboration with teachers and students, to assess the new metrics' usefulness.

References

1. Anohina-Naumeca, A., Grundspenkis, J., Strautmane, M.: The concept map-based assessment system: functional capabilities, evolution, and experimental results. Int. J. Contin. Eng. Educ. Life-Long Learn. **21**(4), 308–327 (2011)
2. Asiksoy, G.: Computer-based concept mapping as a method for enhancing the effectiveness of concept learning in technology-enhanced learning. Sustainability **11**(4), 1005 (2019)
3. Ausubel, D.: The Psychology of Meaningful Verbal Learning. Grane and Stratton, New York (1960)
4. Ausubel, D.P.: Meaningful reception learning and the acquisition of concepts. In: Klausmeier, H.J., Harris, C.W. (eds.) Analyses of Concept Learning, pp. 157–175. Academic Press (1966)
5. Ausubel, D., Novak, J., Hanesian, H.: Educational Psychology: A Cognitive View, 2nd edn. Holt, Rinehart and Winston (1978)
6. Besterfield-Sacre, M., Gerchak, J., Lyons, M., Shuman, L.J., Wolfe, H.: Scoring concept maps: an integrated rubric for assessing engineering education. J. Eng. Educ. **93**(2), 105–115 (2004)
7. Chang, H.C., Chiun-Chieh, H.: Using topic keyword clusters for automatic document clustering. IEICE Trans. Inf. Syst. **88**(8), 1852–1860 (2005)
8. Cliburn, J.: Using concept maps to sequence instructional materials. J. Coll. Sci. Teach. **15**(4), 377–79 (1986)

9. Ferry, B., Hedberg, J., Harper, B.: How do preservice teachers use concept maps to organize their curriculum content knowledge. J. Interact. Learn. Res. **9**, 83–104 (1997)
10. Gove, R.: Gragnostics: fast, interpretable features for comparing graphs. In: The 23rd International Conference Information Visualisation (IV), pp. 201–209 (2019)
11. Himelboim, I., Smith, M.A., Rainie, L., Shneiderman, B., Espina, C.: Classifying Twitter topic-networks using social network analysis. Soc. Media Soc. **3**(1), 1–13 (2017)
12. Jablokow, K., DeFranco, J., Richmond, S.: A statistical study of concept mapping metrics. In: ASEE Annual Conference and Exposition, Conference Proceedings (2013)
13. Kao, G.Y.M., Chen, K.C., Sun, C.T.: Using an e-learning system with integrated concept maps to improve conceptual understanding. Int. J. Instr. Media **37**(2), 151–162 (2010)
14. Kinchin, I., Hay, D., Adams, A.: How a qualitative approach to concept map analysis can be used to aid learning by illustrating patterns of conceptual development. Educ. Res. **42**, 43–57 (2000). https://doi.org/10.1080/001318800363908
15. Leake, D.B., Maguitman, A.G., Cañas, A.J.: Assessing conceptual similarity to support concept mapping. In: FLAIRS Conference, pp. 168–172 (2002)
16. Limongelli, C., Marani, A., Sciarrone, F., Temperini, M.: Measuring the similarity of concept maps according to pedagogical criteria. IEEE Access **10**, 27655–27669 (2022)
17. Marshall, B., Chen, H., Madhusudan, T.: Matching knowledge elements in concept maps using a similarity flooding algorithm. Decis. Support Syst. **42**(3), 1290–1306 (2006)
18. Novak, J.D.: Learning, Creating, and Using Knowledge: Concept Maps as Facilitative Tools in Schools and Corporations. Routledge, NY (2010)
19. Novak, J.: The importance of conceptual schemes for teaching science. Sci. Teach. **31**(6), 10–13 (1964)
20. Novak, J., Gowin, D., Johansen, G.: The use of concept mapping and knowledge Vee mapping with junior high school science students. Sci. Educ. **5**(67), 625–645 (1983)
21. Novak, J.D.: Learning How to Learn. Cambridge University Press, Cambridge (1984)
22. Shannon, C.E.: A mathematical theory of communication. Bell Syst. Tech. J. **27**, 379–423 (1948)
23. Sher, N., Kent, C., Rafaeli, S.: How 'networked' are online collaborative concept-maps? Introducing metrics for quantifying and comparing the 'networkedness' of collaboratively constructed content. Educ. Sci. **10**(10), 267 (2020)
24. Turns, J., Atman, C.J., Adams, R.: Concept maps for engineering education: a cognitively motivated tool supporting varied assessment functions. IEEE Trans. Educ. **43**(2), 164–173 (2000)

Mini-games to Motivate and Engage Users in Learning Recycling Rules

Veronica Rossano(✉), Berardina De Carolis, and Paolodamiano Manzoni

Department of Computer Science, University of Bari, Via Orabona, 4, 70125 Bari, Italy
veronica.rossano@uniba.it

Abstract. Currently, separating waste is one of the widest trend issues in the world to save our planet. However, changing human behaviour is a challenging task that should be started from childhood by using engaging technologies to foster education and change habits. Serious games are one of the main used approaches in education since they have been proved to be engaging and effective in several domains. Recently, also social robots have been implied as an effective technology to motivate students during the learning process. The research described in the paper aims at combining these two technical solutions, serious games and social robots, to educate young people to correctly recycle waste. A user test was conducted to assess the effectiveness of the application, the cognitive workload, and user engagement. Results confirm that children participating in the study improved their recycling skills and the game approach has been judged as engaging.

Keywords: Serious games · Social robots · User experience evaluation

1 Introduction

In September 2015, the United Nation released the 2030 Agenda for Sustainable Development containing 17 goals related to sustainability. The research herein presented addresses one of the goals in the Agenda: to ensure sustainable patterns of production and consumption [1], by raising awareness among people, especially children, of how serious the environmental problem is becoming and how it may affect the quality of our life in the future. Raising awareness in people is a challenging task, changing human behaviours is easier if you educate people from childhood. One of the most effective educational approaches is game-based learning [2, 3] which is used in traditional contexts to promote a better acquisition of both skills and knowledge. Following the evolution of technology, games have been replaced by educational digital games. Recently, the term serious game has been introduced to define videogames or interactive simulations designed and developed not just to entertain [4]. Each serious game has a serious objective, after its use the player will be more aware of a particular topic or will be able to do something [5–7].

In the context of recycling skills, many games have been designed and implemented to train people to classify waste for recycling, since it is one of the most efficient ways to

M. Temperini et al. (Eds.): MIS4TEL 2022, LNNS 580, pp. 75–80, 2023.
https://doi.org/10.1007/978-3-031-20617-7_10

reduce pollution. Some of them are mobile apps or desktop games [8–11], but recently, new interactive approaches have been applied.

However, one of the problems in using serious games in education is the cost of their design and development. Building a video game is very expensive and requires different expertise. However, to be effective a serious game does not have to be complex. Smith and Sanchez [12] described how mini-games can be used to foster conceptual or procedural knowledge. In other words, mini-games are short video games that can be played independently to foster the acquisition of micro-content or specific skills. They are a new methodology for designing and developing serious games, the idea is to set simple tasks to be acquired and to define game elements and mechanics that allow them to be played on their own. In this way, each mini-game has the potential to reinforce a single or small group of learning objectives by providing engaging and motivating experiences. The mini-games approach has been applied in different contexts of education [13, 14] and they have shown to be successful both in terms of gain of relevant knowledge and also in improving collaboration among students. Design and developing mini-games are less expensive both in terms of cost and expertise.

In a previous phase of this research, a serious game has been implemented using the social robot Pepper [15, 16]. PepperRecycle is a game in which the robot and the player compete against each other in a challenge to see who can classify the waste best. To endow the robot with the capability of seeing the waste item and classifying its material correctly, a recognition module based on a Convolutional Neural Network (CNN) that is trained to classify different classes of waste material was developed. The idea is that pupils can practice their skills using the mini-games before being engaged in the challenge against Pepper. To this aim, the two approaches, mobile games, and interaction with a robot have been combined by embedding the mini-games in the social robot Pepper. Thus, the research would evaluate whether the use of mini-games can be useful to engage and train children in specific skills.

The remainder of the paper is organized as follows. In Sect. 2 we introduce the developed mini-games to support recycling skills acquisition. Section 3 describes the outcomes of the pilot study performed for evaluating the effects of the game and the experience in interacting with the social robot. Concluding remarks and future directions are outlined in Sect. 4.

2 Mini-games to Foster Recycling Skills

One of the main problems in waste sorting is to identify the material from which a piece of waste is made of. PeppeRecycle was developed for this purpose. However, user tests [15, 16] highlighted the difficulties of some players who had no prior knowledge about this issue, so the idea was to design a training system to empower children with basic skills. Thus, the PeppeRecycle 2.0 version integrates four mini-games to train players to improve their waste classification skills. The target users are aged between 5 and 12 years old and have a basic knowledge about how to interact with a tablet.

When a game is addressed for learning purposes, one of the goals is to keep the user engaged and motivated to use the game, thus it is important to define mechanics that are tailored to the users' skills. For this reason, the mini-games have been designed at levels of increasing difficulty and to foster knowledge acquisition some microlearning pills [17] have been added. The content could be *hints,* to help the player with the classification, and *trivia,* to explain some details about materials and their danger to the environment. During the design step of mini-games, some inspiration was taken from the applications most widely used by young people to make the interaction familiar to the players. The mini-games are the following:

- *Catch the waste*: objects falling from the sky (Fig. 1a), the player has to move the bin and catch only the right waste;
- *Choose the right bin*: Pepper using its own voice asks the user to answer verbally in which bin the object is the most appropriate to be placed (Fig. 1b);
- *Hit the waste:* the mission is to "hit" the waste of the required type;
- *Space Combat with waste:* the player has to shoot all the rubbish of the wrong category, collecting only the correct ones on the conveyer belt.

In each game, the player can collect points. At the end of each game, Pepper gives some feedback to allow the player to improve their skills and tells some micro knowledge pills. If the player lost all three lives, Pepper would tell some encouraging sentences to push the player to try again. During the game, Pepper's social skills have been applied to enhance the users' experience and motivate the users to play with the robot.

Fig. 1. Screenshots of two mini-games: (a) Catch the waste, (b) Choose the right bin

3 User Study

To measure the effectiveness and the usability of the game a user test has been designed and implemented. Due to the COVID 19 pandemic the test was carried out only on 12 participants. The learning effectiveness was measured using a pre-post test design, the user experience was measured in terms of cognitive workload, using *NASA-TLX* [18], used to measure cognitive load, the *UES (User Engagement Scale)* [19] short-form to measure subjective user engagement, and the *System Usability Scale (SUS)* [20] to measure the user's perceived usability of the games.

3.1 Questionnaires

The pre and post-test questionnaires have been defined using the same elements contained in the game. The two tests are similar and include 4 questions. The first three questions aim at understanding the participants' knowledge about waste recycling before playing the games. The last question aims at evaluating their level of English, since all the buttons in the game are in English, to let the player learn some basic words (play, winner, score, and so forth).

3.2 Sample and Procedure

To test the effectiveness of the game, participants (8 males and 4 females aged between 9 and 10) were asked to play freely to understand what their learning level of waste recycling would be acquired just by playing. As a first step, each participant was asked to fill out the anonymous pre-test questionnaire independently from each other. Then, they played freely with Pepper for about 30 min, and afterwards, they were asked to fill out the post-test questionnaire together with the NASA-TLX, UES and SUS questionnaires.

3.3 Results and Discussion

From the analysis of the collected data, some interesting and encouraging results were obtained. The comparison between pre and post-test questionnaire results indicates that the learning effectiveness of the game has been achieved. The average score of the post-test (7.9) was higher than the one of the pre-tests (6.7), thus the mini-games helped to increase the users' skills. The SUS score was 84/100, therefore according to the general guideline on SUS score, the application is excellent in terms of perceived usability.

Table 1. Results of the NASA-TLX

Cognitive workload measures	Mean	SD
Mental demand	3.7	0.47
Physical demand	1.7	0.94
Temporal demand	1.7	0.94
Effort	6	2.94
Performance	9.3	0.94
Frustration level	2	0

The NASA-TLX results in Table 1 show that some improvements are necessary to make the game less stressful and engaging (Effort 6 out of 10). While the Performance value is a very good result since all the participants were satisfied with their performance. The Frustration measure (2 out 10) means that the users were engaged in the games. Looking at the standard deviation values, there are two interesting points to be discussed. For the Frustration, the value 0 means that all participants expressed the same value, and thus the interaction with Pepper and the mini-games was judged positively. As far as Effort is concerned, the high value means that some participants had some difficulty in playing and some others did not at all.

The UES questionnaire results (Table 2) show that the Focused Attention (2.78 out 5). This value is lower than expected but in line with the Mental Demand in the NASA-TLX. Thus, some improvements in the mini-games mechanics and in the interactions are necessary. However, both the Aesthetic Appeal and Perceived Usability were rated positively. In addition, the values of the standard deviation indicate that all the participants had similar opinions. The Reward (4.56 out of 5) means that all users were satisfied with the experience in the game.

Table 2. Results of UES questionnaire

Engagement measures	Mean	SD
Perceived usability	4	0.16
Focused attention	2.78	0.54
Aesthetic appeal	3.78	0.16
Reward	4.56	0.31

4 Conclusions and Future Works

The data collected on effectiveness is further confirmation of how promising mini-games can be in the educational field. Furthermore, the results obtained for the engagement and cognitive load surveys are satisfactory. Even if the game may seem a bit demanding from a mental point of view, users feel rewarded by the experience. In the future, a storytelling session will be added to exploit the social capabilities of the robot [21]. Moreover, one of the main limitations of the present study was the number of users involved in the test. In the next future, to test the effect of the use of the social robot on the learning process and the user experience, we plan to conduct further studies to confirm these findings and to compare their effectiveness and engagement on a tablet, without the interaction with Pepper.

References s

1. United Nations, Transform our World: The 2030 Agenda for sustainable development. https://sdgs.un.org/2030agenda. Accessed 30 Jan 2022
2. Papastergiou, M.: Digital game-based learning in high school computer science education: impact on educational effectiveness and student motivation. Comput. Educ. **52**(1), 1–12 (2009)
3. Prensky, M.: The Digital Game-Based Learning Revolution. Digital Game-Based Learning. McGraw-Hill, London (2001)
4. Michael, D.R., Chen, S.L.: Serious Games: Games that Educate, Train, and Inform. Muska & Lipman/Premier-Trade, Boston (2005)
5. Hamari, J., Shernoff, D.J., Rowe, E., Coller, B., Asbell-Clarke, J., Edwards, T.: Challenging games help students learn: an empirical study on engagement, flow and immersion in game-based learning. Comput. Hum. Behav. **54**, 170–179 (2016)

6. Bitonto, P., Corriero, N., Pesare, E., Rossano, V., Roselli, T.: Training and learning in e-health using the gamification approach: the trainer interaction. In: Stephanidis, C., Antona, M. (eds.) UAHCI 2014. LNCS, vol. 8515, pp. 228–237. Springer, Cham (2014). https://doi.org/10.1007/978-3-319-07446-7_22
7. Sukajaya, I.N., Mujahidain, D., Suarsana, I.M.: A learning by doing of weight unit conversion-serious game for elementary school students. In: 2020 International Conference on Computer Engineering, Network, and Intelligent Multimedia (CENIM), pp. 168–173. IEEE (2020)
8. Rossano, V., Calvano, G.: Promoting sustainable behavior using serious games: seadventure for ocean literacy. IEEE Access **8**, 196931–196939 (2020)
9. Nunes, E.P., et al.: Mobile serious game proposal for environmental awareness of children. In: 2016 IEEE Frontiers in Education Conference (FIE), pp. 1–8. IEEE (2016)
10. Menon, B.M., Unnikrishnan, R., Muir, A., Bhavani, R.R.: Serious game on recognizing categories of waste, to support a zero-waste recycling program. In: 2017 IEEE 5th International Conference on Serious Games and Applications for Health (SeGAH), pp.1–8 (2017)
11. Prandi, C., Mirri, S., Salomoni, P., Mazzoni, E.: MecWilly in your pocket: on evaluating a mobile serious game for kids. In: 2016 IEEE Symposium on Computers and Communication (ISCC), pp. 185–189. IEEE (2016)
12. Smith, P.A., Sanchez, A.: Mini-games with major impacts. In: Serious game design and development: technologies for training and learning, pp. 1–12. IGI Global (2010)
13. De Jans, S., Van Geit, K., Cauberghe, V., Hudders, L., De Veirman, M.: Using games to raise awareness: how to co-design serious mini-games? Comput. Educ. **110**, 77–87 (2017)
14. Agbo, F.J., Oyelere, S.S., Suhonen, J., Laine, T.H.: Co-design of mini-games for learning computational thinking in an online environment. Educ. Inf. Technol. **26**(5), 5815–5849 (2021)
15. Castellano, G., De Carolis, B., D'Errico, F., Macchiarulo, N., Rossano, V.: PeppeRecycle: improving children's attitude toward recycling by playing with a social robot. Int. J. Soc. Robot. **13**(1), 97–111 (2021). https://doi.org/10.1007/s12369-021-00754-0
16. Castellano, G., De Carolis, B., Macchiarulo, N., Rossano, V.: Learning waste recycling by playing with a social robot. In: 2019 IEEE International Conference on Systems, Man and Cybernetics (SMC), pp. 3805–3810. IEEE (2019)
17. Mohammed, G., Wakil, K., Nawroly, S.: The effectiveness of microlearning to improve students' learning ability. Int. J. Educ. Res. Rev. **3**(3), 32–38 (2018). https://doi.org/10.24331/ijere.415824
18. Hart, S.G.: Nasa-Task Load Index (NASA-TLX); 20 Years Later. In: Human Factors and Ergonomics Society Annual Meeting, vol. 50, no. 9, pp. 904–908 (2006)
19. O'Brien, H.L., Cairns, P., Hall, M.: A practical approach to measuring user engagement with the refined user engagement scale (UES) and new UES short form. Int. J. Hum. Comput. Stud. **112**, 28–39 (2018)
20. Brooke, J.: SUS: a "quick and dirty" usability scale. In: Usability Evaluation in Industry. Taylor and Francis, London (1996)
21. De Carolis, B., D'Errico, F., Rossano, V.: Pepper as a storyteller: exploring the effect of human vs. Robot voice on children's emotional experience. In: Ardito, C., et al. (eds.) Human-Computer Interaction – INTERACT 2021: 18th IFIP TC 13 International Conference, Bari, Italy, August 30 – September 3, 2021, Proceedings, Part II, pp. 471–480. Springer International Publishing, Cham (2021). https://doi.org/10.1007/978-3-030-85616-8_27

Evaluation of the Bibliographical Importance of Digital Educational Disruption Related to Social Networks. The Case of LinkedIn Learning

Javier Parra-Domínguez[1,2](✉), Sergio Manzano[1,2], Andrea Gil-Egido[1,2], Fernando De la Prieta[1,2], Pablo Chamoso[1,2,3], and Sara Rodríguez-González[1,2]

[1] BISITE Research Group, University of Salamanca, Edificio Multiusos I+D+I, C/ Espejos s/n, 37007 Salamanca, Spain
{javierparra,smanzano,age,fer,chamoso,srg}@usal.es

[2] IoT Digital Innovation Hub, Edificio PCUVa, Módulo 117-118, Campus Miguel Delibes Paseo de Belén 9A, 47011 Valladolid, Spain

[3] AIR Institute – Deep Tech Lab, Edificio PCUVa, Módulo 117-118, Campus Miguel Delibes Paseo de Belén 9A, 47011 Valladolid, Spain

Abstract. The social importance of the virtual world is growing. The different social habits have been transferred to digital reality, but issues are as crucial as social relations or education. Concerning the latter, we developed the current research, worrying about how the scientific community can transfer this reality to the different studies that are being developed. In addition to observing the direction of the numerous studies that relate social networks to education, we aim to investigate the evolution of publications associated with a social networking tool such as the one developed by LinkedIn to offer education. As shown, the interest in the possibilities offered by social networks in e-Learning is growing, and the scientific production increases every year, but this does not translate into a significant volume of specific publications on LinkedIn Learning. On the other hand, it is remarkable to observe, in a concrete way, that within the little developed by the scientific community around LinkedIn Learning, 40.91% of these studies relate this social network to the concept of e-Learning.

Keywords: Social networks · e-Learning · LinkedIn Learning

1 Evolution of Education Linked to Digital Development

Education as a fundamental right for all human beings has been with us since 1978 in a constantly evolving society [1]. It is essential to highlight that education development in recent years has been accompanied by the arrival of the Internet [2–4]. Within the evolution close to the Internet, we must highlight the new ways of learning [5–7], highlighting the use of social networks [8–10].

In this article, we will focus on LinkedIn within the aforementioned social networks. In addition to being oriented toward professional and business use [11], business and

M. Temperini et al. (Eds.): MIS4TEL 2022, LNNS 580, pp. 81–86, 2023.
https://doi.org/10.1007/978-3-031-20617-7_11

employment [12], thanks to LinkedIn Learning, the social network offers learning content for its users through courses of interest to them and the possibility of creating and sharing content [13–15]. We are looking for a concrete link with regard to the application of e-learning via LinkedIn, taking into account the boom that has also been brought about by the emergence of the Covid-19 syndemic [16].

Benefits such as e-learning opportunities for students to access academic tutorials, course content and tutors have already been studied [17]. Furthermore, according to Branley and Covey [18] and Awidi et al. [19], social networks are an opportunity for users to interact and share information; hence their intertwining with education is seen as complementary to teaching and learning [20].

This study aims to carry out a bibliometric analysis and research of the evolution of the scientific community's concern regarding new teaching techniques and technologies such as social networks and, more specifically, the advance of the LinkedIn Learning platform.

2 Methodology

To determine the scientific interest aroused by the LinkedIn Learning platform and its importance within the study of social networks as tools dedicated to e-Learning, an analysis of bibliometric indicators was carried out using the Scopus platform. It was decided to use this database because of its importance among the scientific community and the volume of results it could provide. For this same reason, the search equations were formulated comprehensively so that the platform's engine would find on the one hand, all texts in which e-Learning and social networks appeared, and on the other, all texts that related e-Learning and LinkedIn Learning. The search was limited to 2021 to include whole years in the sample and avoid inconsistencies.

For this purpose, a series of equations were designed using query terms connected by the Boolean operators "OR" and "AND," among others.

ALL "E-LEARNING" AND ALL "LINKEDIN LEARNING" PUBYEAR < 2022.

Which yielded 18 results since 2018, when the first publication was recorded - it is important to remember that it was only in 2015 that LinkedIn bought the Lynda platform, which later became LinkedIn Education[1]. One more article has been published this year at the time of writing (April 2022). To narrow down the sample, it was decided to take the year of the first publication in this field as the starting year to analyse its presence compared to social networks in general during the period in which scientific production in this field has been developing. Thus, a second equation was used:

ALL "E-LEARNING" AND ALL "SOCIAL NETWORK" PUBYEAR > 2017 AND PUBYEAR < 2022.

With 7,844 results. This year, to date, 791 texts have been registered in Scopus.

[1] https://news.linkedin.com/2015/linkedin-to-acquire-lyndacom.

Finally, each concept was searched independently to understand the importance of these studies within the total of each idea in general terms. These were the searches carried out and the results they yielded.

ALL "E-LEARNING" PUBYEAR > 2017 AND PUBYEAR < 2022 = 83,580 texts.

ALL "SOCIAL NETWORK" PUBYEAR > 2017 AND PUBYEAR < 2022 = 208,909.

ALL "LINKEDIN LEARNING" PUBYEAR > 2017 AND PUBYEAR < 2022 = 44.

Thanks to the proprietary analysis tool offered by Elsevier on the Scopus website and Excel, the data were analysed. Through this analysis, the researchers aim to answer the following research questions:

- RQ1: What is the trend in the annual volume of publications related to the study of social networking and LinkedIn Learning in e-Learning texts?
- RQ2: Are there differences in the interest in the study of LinkedIn Learning and social networking as an e-Learning tool among the different scientific disciplines?
- RQ3: What percentage of the total number of publications devoted separately to LinkedIn Learning, e-Learning and social media is dedicated to LinkedIn Learning and social media as an e-Learning tool?

In this way, the aim is to find out how much interest this type of research awakens in academic authors and whether it has increased over time, within which fields of study there is more scientific production in this respect, and what its presence is within the total scientific output of each field understood independently.

3 Results

As can be seen in Figs. 1 and 2, although the trend is linear in both cases, both the volume of annual publications and the annual difference, upwards, is considerably higher in the case of texts related to e-Learning and social networks. However, the volume of publications on e-Learning and LinkedIn Learning is still too low to draw firm conclusions. This answers to RQ1. To answer RQ2, Table 1 is presented, which shows the five main fields of study from which both issues have been addressed. While the main subject of the

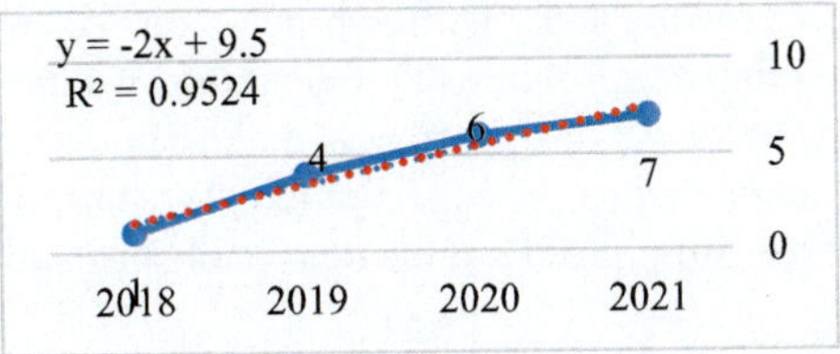

Fig. 1. Documents published by year and trend in *e-Learning* and *LinkedIn Learning*

texts in e-Learning and LinkedIn Learning is Social Sciences (31.25%), in e-Learning and social networks, it is Computer Science (30%).

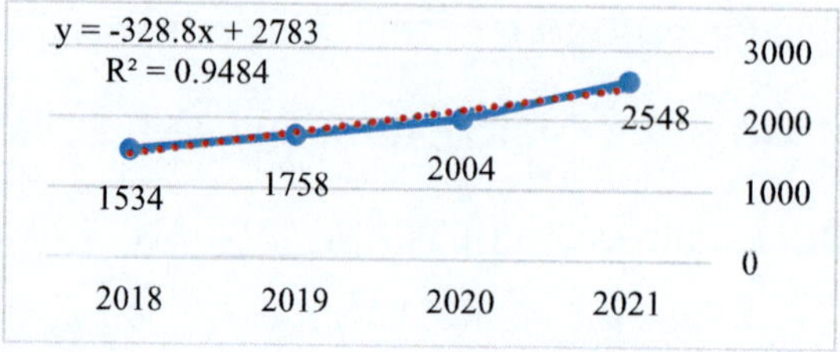

Fig. 2. Documents published by year and trend in *e-Learning* and *social network*

Table 1. Relating each search to the isolated search of each concept

	E-LEARNING AND *LINKEDIN LEARNING*	*E-LEARNING* AND *SOCIAL NETWORK*
Social Sciences	**31.25%**	22.96%
Computer Science	18.75%	**30%**
Decision Sciences	9.38%	
Engineering	9.38%	10.78%
BUSI, MGMT & Accounting	6.25%	7.11%
Mathemathics		4.81%

Finally, in an effort to respond RQ3, the relationship between each item and its more general search has been sought. Within the texts in which *LinkedIn Learning* appears, the studies that mention *e-Learning* account for 40.91%, while of the total number of studies in which *e-Learning* appears, *LinkedIn Learning* only occurs in 0.02%.

In the second case, 9.39% of the studies on *e-Learning* also include social networks, but 3.75% of the works on networks have a reference to *e-Learning*.

4 Discussion and Conclusions

As can be seen from RQ1, although interest in the possibilities offered by social networks concerning e-Learning is growing and scientific production is increasing every year, this does not translate into a significant volume of publications specific to LinkedIn Learning on the Scopus platform. Moreover, this interest is mainly in the branches of knowledge directly related to the study change, as expected (RQ2). Thus, in both circumstances, although in a different order, around 50% of the texts published are in the Social Sciences or Computer Science.

It is remarkable what percentage of these types of studies are represented within their broader field (RQ3). It is particularly striking that in almost 10% of the studies on e-Learning, there is a presence of content related to social networks; and it is also curious to note how, of the few articles that have been written on LinkedIn Learning, 40.91% of them refer to the concept of e-Learning. This is a relatively low percentage considering this is essentially what this platform is based on.

Despite the methodological limitations of this study, it is clear that academia is increasingly focusing on the opportunities offered by e-Learning through the use of social networks. This is not the case for LinkedIn Learning. The volume of publications on e-Learning and LinkedIn Learning is still too low to draw firm conclusions, so it is necessary to develop a series of case studies with this tool, linking it to other platforms and other social networks, demonstrating its usefulness and importance within this discipline. Therefore, it would be interesting to extend the study with the analysis of different databases, in order to cover a larger and more significant sample and compare the results with the presence of these studies in the platform that concerns the present analysis.

Likewise, in both cases, both in the case of social networks and e-Learning in general, and LinkedIn Learning in particular, it would be interesting to extend the study by taking into account other variables such as the prominent publications that publish these studies, their quality and impact indices, or this research could even be carried out through a content analysis that would study in detail the approach taken to the subject in the different publications. Nevertheless, as we have seen in the theoretical framework, it is evident that e-Learning and social networks have become an essential pillar of education today, so it is crucial to carry out quality studies that analyse the opportunities, risks or training methodologies that can help to develop.

To conclude, it is clear from the results presented that a new trend is emerging in training processes and higher education should not remain unaffected. It is necessary, on the one hand, that academic institutions understand these platforms as an opportunity to diversify their offer and the formats in which they are offered, and on the other hand, it is necessary to train professionals capable of managing these platforms and offer content that provides a differential value to traditional training.

Acknowledgments. This work has been partially supported by the European Regional Development Fund (ERDF) through the Interreg Spain-Portugal V – A Program (POCTEP) under gran 0677_DISRUPTIVE_2_E (Intensifying the activity of Digital Innovation Hubs within the PocTep region to boost the development of disruptive and last generation ICTs through cross-border cooperation).

References

1. McCowan, T.: Reframing the universal right to education. Comp. Educ. **46**(4), 509–525 (2010)
2. Mokhtari, K., Reichard, C.A., Gardner, A.: The impact of internet and television use on the reading habits and practices of college students. J. Adolesc. Health. **52**(7), 609–619 (2009)
3. Miguel Tomé, S.: Towards a model-theoretic framework for describing the semantic aspects of cognitive processes. ADCAIJ: Adv. Distrib. Comput. Artif. Intell. J. **8**(4), 83–96 (2020)
4. Ivanova, M., Petrova, T.: Analysis of relationship between students' creative skill and learning performance. In: Kubincová, Z., Lancia, L., Popescu, E., Nakayama, M., Scarano, V., Gil, A.B. (eds.) MIS4TEL 2020. AISC, vol. 1236, pp. 66–75. Springer, Cham (2020). https://doi.org/10.1007/978-3-030-52287-2_7

5. Aljojo, N., et al.: Kids' atlas application to learn about geography and maps. ADCAIJ: Adv. Distrib. Comput. Artif. Intell. J. **9**(2), 33–48 (2020)
6. Vergara, D., Extremera, J., Rubio, M.P., Dávila, L.P.: The proliferation of virtual laboratories in educational fields. ADCAIJ: Adv. Distrib. Comput. Artif. Intell. J. **9**(1), 85–97 (2020)
7. Nakayama, M., Sciarrone, F., Uto, M., Temperini, M.: Estimating student's performance based on item response theory in a MOOC environment with peer assessment. In: Kubincová, Z., Lancia, L., Popescu, E., Nakayama, M., Scarano, V., Gil, A.B. (eds.) MIS4TEL 2020. AISC, vol. 1236, pp. 25–35. Springer, Cham (2020). https://doi.org/10.1007/978-3-030-52287-2_3
8. Buxarrais, M.R.: Social networks and education. Redes sociales y educación, 15–20 (2016)
9. Addone, A., Palmieri, G., Pellegrino, M.A.: Engaging children in digital storytelling. In: De la Prieta, F., et al. (eds.) MIS4TEL 2021. LNNS, vol. 326, pp. 261–270. Springer, Cham (2022). https://doi.org/10.1007/978-3-030-86618-1_26
10. Salazar, O., Álvarez, S., Ovalle, D.: EoLo: a serious mobile game to support learning processes. Adv. Intell. Syst. Comput. **617**, 118–125 (2017)
11. Parra, J., Pérez, E., Rodríguez, J.M.C.: Capítulo VI. Estudio de las diferencias de género en la comunicación a través de las redes sociales. El caso de la utilización de linkedin como palanca de transferencia de conocimiento por parte de los equipos de gobierno universitarios de las universidades públicas españolas. Nuevas pantallas para la comunicación digital: Redes sociales, infoentretenimiento y consumo creativo del audiovisual. Ed. Egregius (2021)
12. Koch, T., Gerber, C., De Klerk, J.J.: The impact of social media on recruitment: are you LinkedIn? J. Hum. Resour. Manag. **16**(1), 1–14 (2018)
13. ten Brink, M., Bredeweg, B., Schouten, B.: Students' attitude towards technology enhanced photovoice to support critical reflection. In: De la Prieta, F., et al. (eds.) MIS4TEL 2021. LNNS, vol. 326, pp. 32–41. Springer, Cham (2022). https://doi.org/10.1007/978-3-030-86618-1_4
14. Du, X.: Embedding LinkedIn learning MOOCs to enhance student's educational experience and employability. In: European Conference on e-Learning, 163 – XIV. Academic Conferences International Limited (2020)
15. Leon, F., Popescu, E.: Using large margin nearest neighbor regression algorithm to predict student grades based on social media traces. Adv. Intell. Syst. Comput. **617**, 12–19 (2017)
16. Verma, A.K., Prakash, S.: Impact of COVID-19 on environment and society. J. Glob. Biosci. **9**(5), 7352–7363 (2020)
17. Gikas, J., Grant, M.M.: Mobile computing devices in higher education: student perspectives on learning with cellphones, smartphones & social media. Internet High. Educ. **19**, 18–26 (2013)
18. Branley, D.B., Covey, J.: Risky behavior via social media: the role of reasoned and social reactive pathways. Comput. Hum. Behav. **78**, 183–191 (2018)
19. Awidi, I.T., Paynter, M., Vujosevic, T.: Facebook group in the learning design of a higher education course: an analysis of factors influencing positive learning experience for students. Comput. Educ. **129**, 106–121 (2019)
20. Shodiyev, M.B.: The usage of web technologies as social network (Facebook) in teaching a foreign language to adults. Sci. Educ. **3**(2), 973–977 (2022)

May a Distance Learning Course in Statistics Satisfy Medical Students? The Experience with an Italian University Sample During the Covid Pandemic

Vincenza Cofini(✉), Mario Muselli, Pierpaolo Vittorini, Annalucia Moretti, and Stefano Necozione

Department of Life, Health and Environmental Sciences, University of L'Aquila, P.le S. Tommasi, 1 - Coppito, 67100 L'Aquila, Italy
vincenza.cofini@cc.univaq.it, {mario.muselli, annalucia.moretti}@graduate.univaq.it, {pierpaolo.vittorini, stefano.necozione}@univaq.it

Abstract. Teaching medical statistics online poses several difficulties related both to the subject and the use of technology. Due to the Covid-19 pandemic, a traditional medical statistics course was forced to be held online. The study investigates the impact of such a transformation in terms of both the student satisfaction and the stress at the end of the course, by taking into account the attitudes toward statistics before and after the academic course. The results show that the students' learning satisfaction was on average 30.1 (5.8). Analyzing the singles items: over 70% of students were satisfied of the technology used, even if only about 18% would participate in a new online course. Students were stressed and half of them reported high levels of stress.

Keywords: TEL in medical students · Medical statistics learning · Stress · Statistics attitude

1 Introduction

The curriculum of the degree course in Medicine and Surgery at the University of L'Aquila (Italy), for the Academic Year 2020–21, included in the first year the module called "Medical Statistics".

In October 2020, only the first lesson of the module was conducted traditionally, in the classroom. Then, due to the Covid pandemic, the traditional teaching was substituted by a distance learning approach. The lessons were not interrupted and both the teacher and the students had to re-organize the teaching and learning activities to satisfy the new approach, for the whole time of the course. The module was then offered using Microsoft Teams for video conferences, Moodle to deliver the learning material and for the final evaluation.

M. Temperini et al. (Eds.): MIS4TEL 2022, LNNS 580, pp. 87–96, 2023.
https://doi.org/10.1007/978-3-031-20617-7_12

Such a sudden recourse to distance learning may have had an impact on the quality of teaching and therefore on the understanding of the lectures. Accordingly, the aim of the present study is to assess the impact of the forced transformation of the traditional course into a distance learning one, by investigating both the student satisfaction and the stress at the end of the course, taking into account the attitudes toward statistics before and after the academic course.

The paper is organized as follows. Section 2 is dedicated to the related work. Then, Sect. 3 describes the module structure in detail. Section 4 focuses on describing the study research, whereas Sect. 5 reports the results. Finally, Sect. 6 ends the paper with a discussion about the most relevant findings.

2 Related Work

Several approaches exist to improve statistics learning [1]. As reported in literature, both theory and application in medical statistics should be introduced with clinical examples and this should extend to the choice of data type and dataset for the analysis [2].

In Italy, the teaching approach of medical statistics is usually oriented on the Evidence Based Medicine concept (EBM) [3]. Students have to be trained on the importance and role of statistics in scientific medical development from the planning of a clinical study to the choice of methods functional to the objectives, to the analysis of the data and to the interpretation of the results.

Teaching statistics in a medical course requires an important effort by the teacher to bring students closer to a subject that they may have only marginally learned at school. A recent study has reported that medical students have negative attitudes toward statistics, and many experienced anxiety [4]. Another study evidenced that Medical postgraduates showed positive attitudes toward statistics, but they reported that statistics was a very difficult subject [5].

Another fact to taking into account is that medical students are particularly stressed by the enormous commitment they are put through. Studies carried out before the Covid pandemic highlighted the psychological distress that medical students suffered [6, 7]. Furthermore, several studies have reported the negative impact of Covid pandemic on the quality of life of students who have experienced psychological distress related to the fear of infection but also to social distancing [8].

As with any subject, understanding the difficulties, fears and efforts of students can help the teacher to choose the best approach satisfactory for all concerned.

The use of technology is useful to actively modulating the understanding of the methods introduced and the active participation of students and this was experimented for specific fields like medical statistics [9]. Scientific evidence has highlighted how the use of technology helps comprehension in situations of particular difficulty [10]: it is therefore possible that its use in the statistical field can help understanding the subject, also in a singular and stressful situation such as the pandemic. Several authors have tried to understand the best approaches to teach statistics online [11–13], defining effective practices that might be useful to teachers teaching statistics online.

A recent study by the Education Committee of the Italian Society of Medical Statistics and Clinical Epidemiology, investigated the teachers' perceptions about online teaching of medical statistics, evidencing that 61% of Italian academics of medical statistics reported a positive experience and they were favorable to provide online teaching of medical statistics, as well as biostatistic and epidemiology in the future. Authors warned that the teacher's presence in the classroom and the human interaction remain fundamental components to learning and it's necessary to know the student's satisfaction [14].

3 Module Structure

3.1 Medical Statistics Module

The module was scheduled to take place from October 2020 to January 2021 at three hours per week for a total amount of 37.5 h, with the goal to introduce learners to the basic statistical methods and its application to understand biomedical scientific research. Students should comprehend descriptive statistics, understand the basic inferential methods and demonstrate capacity for reading and understand the results of the statistical analysis in biomedical research. The teaching methods included the combination of traditional teaching for lectures and problem-solving classes and a blended approach, with homework exercises offered through the free Moodle learning platform available at https://moodle.univaq.it/ and it was introduced during the first lesson [9].

The module's subjects were:

- Descriptive statistics - Data types, summary statistics, tables and graphs, simple linear regression, correlation, Bayes Theorem;
- Inferential Statistics - Probability distribution, technical sampling, Confidence intervals and statistics test: z test, t-test, X^2-test, One Way Analysis of Variance.

After the first lesson, the course was fully online, lecturers were through the Microsoft Teams platform. A team called "Medical Statistics" was created to video conference, the time of lesson was reduced to 40 min then the teacher organized many activities for virtual classroom exercises and homework exercises using both Microsoft teams and Moodle platform [15]. In particular, Moodle was chosen by our University as the supporting platform, to both deliver learning material and to perform online assessments. As for the online tests, the University of L'Aquila suggested the adoption of the Safe Exam Browser as proctoring system [16].

During the course, each lesson was re-organized as follow:

- First part: Introduction to theoretical content, illustrating the topics with the help of published medical research or research's protocols:
- Second part: Application on data from available dataset;
- Third part: self-evaluation test.

At the end of each part, students had to solve problems and quiz on Moodle platform for self-assessment.

To introduce the role of statistics in scientific medical research, during Covid pandemic era, were presented studies related to Sars-Cov 2 infection, with particular reference to the treatment of the disease and epidemiological methods for monitoring the spread of the epidemic. Covid pandemic was the opportunity to work with students with epidemiological data collected and analyzed during the last year by our research group [17, 18]. To introduce the randomization concept and inferential methods we used data from randomized controlled clinical trials carried out in the University Clinical Unit of San Salvatore Hospital [19, 20].

At the end of the program's module, the students were evaluated with a test lasting 30 min, whose items and related answers were randomly selected from the Moodle repository and weighted (in terms of their difficulty) by the teacher.

4 The Study

All the students admitted at the Medicine and Surgery degree in the Academic Year 2020/2021 were invited to fill an anonymous questionnaire both pre and post-course.

All students were informed by the teacher about the project objectives and they participated voluntarily after giving their informed consent.

We used the following instruments:

1. the "Survey of Attitudes Toward Statistics" questionnaire (SATS-36) [21, 22]: this questionnaire evaluates the students' attitudes for statistic. It contains 36 items that explore 6 components. 4 components included in SATS-28 and 2 components added in the SATS-36, the first 4 components are Affect, Cognitive Competence, Value, and Difficulty and the last 2 components are Interest and Effort. "Affect" (6 items) investigates the feelings relating to statistics, "Cognitive Competence" (6 items) investigates students' attitudes to applying to statistics their intellectual knowledge and skills, "Value" (9 items) investigates about the usefulness, relevance, and worth of statistics in personal and professional life, "Difficulty" (7 items) investigates the difficulty of statistics as a subject, "Interest" (4 items) evaluates individual interest in statistics and, finally, "Effort" (4 items) estimates the fatigue the student employs to study statistics. Each item has 7 selections allowed in a range from 1 (strongly disagree) to 7 (strongly agree). Finally some items ask for relevant demographic and academic background information. Students completed the questionnaire twice: before and post the statistical course, with some differences about following items. The SATS instruments and scoring guides are available at http://www.evaluationandstatistics.com and the copyright is hold by Candace Schau [22].

   In this paper we reported also the results about the grade point average estimated by students and the answers to three single global attitude items from the SATS-36 pre and post course questionnaire evaluated:

a. Math cognitive competence, with the item: "*How good at mathematics are you?*" [scoring from 1 to 7 (very poor to very good)];
b. Career Value, with the item: " I*n the field in which you hope to be employed when you finish school, how much will you use statistics?*" [scoring from 1 to 7 (Not All- Great Deal)]
c. Statistics Cognitive Competence. With the item: "*How confident are you that you have mastered introductory statistics material?*" [scoring from 1 to 7 (Not at all confident - Very confident)]

At end of the course we assessed the satisfaction of the online course and the stress perceived with the following questionnaires:

2. The "Satisfaction Scale" [23]: this questionnaire consists of 10 items scored on a Likert scale of 1–5 with a score of 1 = strongly disagree, 2 = disagree, 3 = uncertain, 4-agree, and 5 = strongly agree. We used an adapted version of it.
3. the "GH-12 questionnaire" [24]: this questionnaire is used to assess perceived psychological distress and consists of 12 items that assess the severity of a mental problem over the past few weeks using a 4-point scale (from 0 to 3). A GH-12 score ≤ 15 indicates an average stress level, a 15–20 score indicates a moderate level of stress, and a score ≥ 20 indicates more intense psychological distress.

All variables were analyzed, and quantitative data were reported as mean and standard deviation (SD) or median and range while categorical data as absolute frequencies and percentages. Chi square test and Wilcoxon matched-pairs signed-ranks test were run for matched data comparisons. To compare repeated measures the Friedman test was used. All analyses were performed using STATA 14, setting $p < 0.05$ as the threshold for statistical significance.

5 Results

One hundred and seventy students registered on the Microsoft Teams and one hundred and seventy-three on the Moodle platform, 123 complied with the minimum attendance (as required for the Degree in Medicine and Surgery in Italy) and took the exam. The SATS-36 pre course questionnaire was administered to 130/173 subjects but in the second survey, at the end of the lessons and before the final test only in 64/123 fully compiled the Sat test, 54/64 and 52/64 students completed the remaining questionnaires respectively (Satisfaction scale and GH-12 questionnaire). The present study reports the analysis for matched data. They were prevalent females 41 (64%) with a mean age of 19 years (2.4). With respect to Math cognitive competence, 2 students reported a very poor level (3%) and six a very good level (9%), the mean and median values were 4.8 (1.5) and 5 (1–7). Same mean and median values were reported for the other two items used to investigate Career Value and Statistics Cognitive competence (Fig. 1).

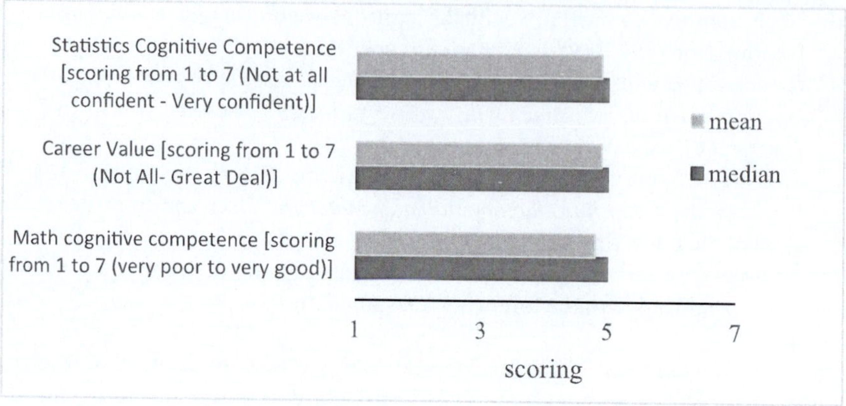

Fig. 1. Global attitude items of participants

Table 1 reports the comparison between pre and post scores of the Attitude components of the SATS-36 questionnaire. Only for the Effort component there was a significant difference ($p < 0.001$), indicating that the amount of work the student expended to learn statistics post course was lower than expected.

Table 1. Pre-course/post course SATS-36 Test

Attitude components	Pre mean (SD) median (range)	Post mean (SD) median (range)	p-value
Affect	4.43 (0.95) 4.33 (1.83–6.33)	4.45 (0.93) 4.50 (2.50–6.17)	0.765
Cognitive competence	5.03 (0.76) 5.00 (2.83–6.33)	4.99 (0.65) 5.00 (3.50–6.33)	0.571
Value	5.06 (0.73) 5.11 (3.11–6.33)	4.96 (0.71) 5.00 (3.56–6.33)	0.220
Difficulty	2.89 (0.65) 2.86 (1.29–4.14)	2.86 (0.71) 2.86 (1.29–4.43)	0.656
Interest	5.87 (0.77) 6.00 (4.00–7.00)	5.95 (0.79) 6.00 (4.00–7.00)	0.929
Effort	6.36 (0.65) 6.50 ((3.74–7.00)	5.84 (0.89) 6.00 (3.00–7.00)	< 0.001

With respect to satisfaction of learning at distance, as reported in Table 2, only 1 student was strongly unsatisfied, and five students were unsatisfied with technology used during the course. Students suffered from difficulty concentrating and only 15 of them reported that "*Online lessons facilitated the learning of the topics of the course*".

Overall mean score on the learning satisfaction was 30.1 (5.8) with a median value of 30.5 (18–42).

Table 2. Satisfaction scale Items: mean (SD).

Items	SD	D	U	A	SA
I was able to learn from the online lessons	4 (7, 41)	4 (7, 41)	5 (9, 26)	25 (46, 3)	16 (29, 63
I was stimulated to do additional reading or research on topics discussed	7 (12, 96)	16 (29, 63)	14 (25, 93	9 (16, 67)	8 (14, 81)
I would like to participate in another online course in the future	22 (40, 74)	12 (22, 22)	10 (18, 52)	3 (5, 56)	7 (12, 96)
I am satisfied with the technology used in this course	1 (1, 85)	5 (9, 26)	9 (16, 67)	26 (48, 15)	13 (24, 07)
The diversity of topics in the online course prompted me to participate in the discussions	10 (18, 52)	13 (24, 07)	14 (25, 93)	13 (24, 07)	4 (7, 41)
I had difficulty learning how to use the platform	25 (46, 30)	23 (42, 59)	4 (7, 41)	1 (1, 85)	1 (1, 85)
I enjoyed participating in this course	11 (20, 37)	6 (11, 11)	19 (35, 19)	10 (18, 52)	8 (14, 81)
Professor was available and helpful in facilitating the use of the platform	–	2 (3, 70)	3 (5, 56)	30 (55, 56)	19 (35, 19)
I had difficulty concentrating	11 (20, 37)	15 (27, 78)	10 (18, 52)	13 (24, 07)	5 (9, 26)
Online lessons facilitated the learning of the topics of the course	6 (11, 11)	17 (31, 48)	16 (29, 63)	9 (16, 67)	6 (11, 11)

Note: SD = Strongly disagree; D = Disagree; U = Uncertain; A = Agree, SA = Strongly agree.

After the end of the course, the mean score of stress was 20.4 (6.7) indicating the presence of intense psychological distress. Half of the students (26/52) reported scores higher than 20 (50%) and 13 students (25%) reported a moderate stress (score: 15–19).

Before and after the course, students were asked "What grade do you expect to receive in this course?". The mean of the final grade, as reported in Table 3, was higher than the expected grade even if there were no significant differences (Friedmann chi square = 0.376; p = 0.829).

Table 3. Expected and observed grade point average

	n	Mean (SD)
Precourse grade expected	64	27, 7 (2, 7)
Post course grade expected	64	28, 1 (2, 5)
Final grade	60	28, 2 (2, 9)

(Note: at the time of writing, 4 students did not take the exam yet).

6 Discussion and Conclusions

The aim of the study was to assess the student's satisfaction and stress perceived during the online course of medical statistics, assessing their attitudes towards statistics pre and post course.

Students reported a decrement in the Effort component and all other investigated components remained stable over the time. The learning satisfaction was discrete [30.1 (5.8)]; over 70% of students were satisfied of the technology used, even if only about 18% would participate in a new online future course. It is worth remarking that, instead of using general tools for measuring satisfaction (e.g., System Usability Scale [25]), we used a survey instrument specifically developed and validated for online courses [23], thus strengthening our findings.

As expected, students appear stressed (75%) and half of them reported high levels of stress. Nevertheless, a pre-pandemic study on 2455 students which also enrolled medical students from University of L'Aquila, reported that 72% of participants were at risk of perceived stress with different distribution by categories (medium risk score: 55.2% and "high-risk" score: 16.9%) [26]. Even if it was measured with different instruments and on students in 1st, 4th, and 6th year of the course, our findings do not suggest an increased stress on students. This study does not allow us to establish whether the stress detected is due to distance learning, for various reasons, we do not know other variables that can influence stress and above all the study design does not allow us to know the causal relationship.

Furthermore, a recent study [27] on medical students from the same University described the effect of the transformation of a blended learning course regarding data science into a fully online course, due to the pandemic. Differently from our study, the authors already used in the previous years a web site to deliver the learning material and an ad-hoc platform to support formative and summative assessment [28]. The results surprisingly showed - with respect to the previous academic year - an increased perceived didactic quality, engagement and didactic outcomes (even if the latter not statistically significant).

To conclude, all our findings seem to be in line with the literature, and - even if our study has some limitations like the lack of randomization, students seem to be satisfied with the active participation and stimulation offered by the adopted technologies, they reported that technologies could help them in learning statistics (as in [29]), on the other hand most of them would not prefer the online course in the future.

References

1. Tishkovskaya, S., Lancaster, G.A.: Statistical education in the 21st century: a review of challenges, teaching innovations and strategies for reform. J. Stat. Educ. **20**, 2 (2012)
2. MacDougall, M., Cameron, H.S., Maxwell, S.R.J.: Medical graduate views on statistical learning needs for clinical practice: a comprehensive survey. BMC Med. Educ. **20**(1), 1 (2019)
3. Morris, R.W.: Does EBM offer the best opportunity yet for teaching medical statistics? Stat. Med. **21**(7), 969–977 (2002). discussion 979–981, 983–984
4. Althubaiti, A.: Attitudes of medical students toward statistics in medical research: evidence from Saudi Arabia. J. Stat. Data Sci. Educ. **29**(1), 115–121 (2021)
5. Zhang, Y., et al.: Attitudes toward statistics in medical postgraduates: measuring, evaluating and monitoring. BMC Med. Educ. **12**, 117 (2012)
6. Bert, F., et al.: Prevalence of depressive symptoms among Italian medical students: the multicentre cross-sectional "PRIMES" study. PLoS One **15**(4), e0231845 (2020)
7. Abdulghani, H.M., AlKanhal, A.A., Mahmoud, E.S., Ponnamperuma, G.G., Alfaris, E.A.: Stress and its effects on medical students: a cross-sectional study at a college of medicine in Saudi Arabia. J. Health Popul. Nutr. **29**(5), 516–522 (2011)
8. Charbonnier, E., Goncalves, A., Puechlong, C., Montalescot, L., Le Vigouroux, S.: Two years and four time points: description of emotional state and coping strategies of french university students during the COVID-19 pandemic. Viruses **14**, 782 (2022)
9. Cofini, V., Vittorini, P.: Evaluating statistical and informatics competencies in medical students in a blended learning course. In: Kubincová, Z., Lancia, L., Popescu, E., Nakayama, M., Scarano, V., Gil, A.B. (eds.) MIS4TEL 2020. AISC, vol. 1236, pp. 76–85. Springer, Cham (2021). https://doi.org/10.1007/978-3-030-52287-2_8
10. Cofini, V., Di Giacomo, D., Di Mascio, T., Necozione, S., Vittorini, P.: Evaluation plan of TERENCE: when the user-centred design meets the evidence-based approach In: International Workshop on Evidence-Based Technology Enhanced Learning, 152, AISC Advances in Intelligent and Soft Computing, pp. 11-18, Springer, Berlin (2012). https://doi.org/10.1007/978-3-642-28801-2_2
11. Jamie, D.M., Dheeraj, R.: Teaching statistics online: a decade's review of the literature about what works. J. Stat. Educ. **19**(2) (2011)
12. Monnie, M.G., Lynne, S., Pavel, N.: Just-in-time teaching in statistics classrooms. J. Stat. Educ. **24**(1), 16–26 (2016)
13. Haardörfer, R., Livingston, M.: Teaching statistics in public health during COVID-19: lessons learned from spring 2020 and recommendations for the future. Pedagogy Health Promot. **7**(1), 25–28 (2021)
14. Rota, M., Peveri, G., Fanelli, M., Torelli, L., Rocchi, M.B., Specchia, C.: Satisfaction with online teaching of medical statistics during the COVID-19 pandemic: a survey by the education committee of the Italian society of medical statistics and clinical epidemiology. Teach. Stat. **43**(3), 129–139 (2021)
15. Florjancic, V., Wiechetek, L.: Using moodle and MS teams in higher education - a comparative study. Int. J. Innov. Learn. **31**(2), 264–286 (2022)
16. Arnò, S., Galassi, A., Tommasi, M., Saggino, A., Vittorini, P.: State-of-the-art of commercial proctoring systems and their use in academic online exams. IJDET **19**(2), 55–76 (2021)
17. Muselli, M., Cofini, V., Desideri, G., Necozione, S.: Coronavirus (Covid-19) pandemic: how may communication strategies influence our behaviours? Int. J. Disaster Risk Reduct. **53**, 101982 (2021)
18. Picchi, G., et al.: Laboratory parameters related to severe disease and death in SARS-CoV-2 pneumonia: retrospective analysis. J Med Virol. **93**(10), 5886–5895 (2021)

19. Fusco, P., et al.: Unilateral paravertebral block compared with subarachnoid anesthesia for the management of postoperative pain syndrome after inguinal herniorrhaphy: a randomized controlled clinical trial. Pain **157**(5), 1105–1113 (2016)
20. Fusco, P., et al.: Continuous wound infusion and local infiltration analgesia for postoperative pain and rehabilitation after total hip arthroplasty. Minerva Anestesiol. **84**(5), 556–564 (2018)
21. Schau, C., Stevens, J., Dauphinee, T.L., Del Vecchio, A.: The development and validation of the survey of attitudes toward statistics. Educ. Psychol. Measur. **55**, 868–875 (1995)
22. Schau C.: CS Consultants, LLC (2019). https://www.evaluationandstatistics.com/. Accessed 08 May 2022
23. Gunawardena, C.N., Zittle, F.J.: Social presence as a predictor of satisfaction with a computer-mediated conferencing environment. Am. J. Distance Educ. **11**, 8–26 (1997)
24. Goldberg, D.P., Williams, P.A.: User's guide to the general health questionnaire, 1st ed., NFER-Nelson, Windsor, Canada (1998)
25. Brooke, J.: SUS: a "quick and dirty" usability scale. Usability evaluation in industry. (1996)
26. Leombruni, P., et al.: Stress in Medical students: PRIMES, an Italian, multicenter cross-sectional study. Int. J. Environ. Res. Public Health **19**, 5010 (2022)
27. Vittorini, P., Galassi, A.: From blended to online due to the COVID outbreak: the case study of a data science course. Open Learn.: J. Open Distance e-Learn. **36**(3), 212–230 (2021)
28. Bernardi, A., et al.: On the design and development of an assessment system with adaptive capabilities. In: Di Mascio, T., et al. (eds.) MIS4TEL 2018. AISC, vol. 804, pp. 190–199. Springer, Cham (2019). https://doi.org/10.1007/978-3-319-98872-6_23
29. Shariati, B., Kanji, Z., Soheilipour, S., Patrick, L., Sharif, A.: Enhancing learning in an online oral epidemiology and statistics course. Can. J. Dent. Hyg. **55**(1), 17–29 (2021)

Combining Learner Model and Reinforcement Learning for Adaptive Sequencing of Learning Activities

Amel Yessad(✉)

Sorbonne Université, CNRS, LIP6, 4 Place Jussieu, 75252 Paris Cedex 05, France
amel.yessad@lip6.fr

Abstract. In this paper, we present an approach for adapting the sequencing of learning activities that relies on the Q-learning, a reinforcement learning algorithm. The Q-learning learns a sequencing policy to select learning activities that improves the knowledge states of students.

In this research, we rely on the student knowledge state inferred by the Bayesian Knowledge Tracing (BKT) at every testing activity to calculate the reward of the Q-Learning. The more the Q-Learning decision improves the student knowledge state the greater the reward received by the Q-Learning. In addition, we propose a 3-step method aiming to ensure that the use of the Q-Learning is education domain compliant. It consists on training the Q-Learning first on simulated students to answer the "cold start" problem of the Q-Learning.

We present empirical results showing that the sequencing policy resulting from the 3-step method provides the ITS with an efficient strategy to improve the students' knowledge states.

Keywords: Adaptive instruction · Reinforcement learning · Q-Learning · Bayesian knowledge tracing

1 Introduction

In intelligent tutoring systems (ITS), Curriculum Sequencing has been widely studied and consists on the planned sequence of learning activities (definitions, examples, questions, problems, etc.) that are most suitable according to the student caracteristics [1,4].

Several research [2,5,6] have shown the interest of reinforcement learning (RL) for instructional policies, as RL models can learn complex and latent relationships between instructional tasks, students actions, and knowledge outcomes. In particular, the problem of sequencing the learning activities in ITS according to the student characteristics fits well a RL problem [2]. RL models need to be trained on student historical data to converge to a good sequencing policy for ITS [7]. However, in education, policy evaluation becomes challenging: experience is scarce and expensive, and the human mind is part of the environment of

M. Temperini et al. (Eds.): MIS4TEL 2022, LNNS 580, pp. 97–102, 2023.
https://doi.org/10.1007/978-3-031-20617-7_13

the RL agent [7]. In this paper, we propose a method of three steps to answer the issue of data availability and we carry out an initial evaluation of this method.

In this research, we tackle the issue of the adaptive sequencing of learning activities by combining the Q-learning algorithm [8] and the BKT student model [3].

We first formalise the problem of sequencing of learning activities by defining the main elements of the Q-Learning: the Q-table and the reward function. Then, we propose to initialize the Q-table with simulated students in order to answer the "cold start" problem of the Q-Learning, to fast its convergence and maximize the student learning gain. These simulated data were generated with the support of experts who defined rules about the students' behaviours.

The paper is organized as follows: first, the principle of the Q-Learning is presented. The proposed approach and our contributions are summarized in Sect. 3. Then, the learning process and the experimental study are described in Sect. 5. Finally, the main conclusions are given in Sect. 7.

2 Q-Learning: Q-Table and Q-Function

Q-Learning is a RL algorithm where an agent learns the make decisions (actions) in different situations (states) through trial and error. It is relies on (1) a table, named Q-table that associates observed states s with actions a and (2) a function, named Q-function that maximizes a "value function" $Q(s,a)$ of an action a for a state s.

In our case, the Q-table is the data structure used to calculate the maximum expected future rewards for each learning activity at each student knowledge state. This table will guide the Q-Learning agent to select the "best" learning activity for each student to maximize her learning gain. Each value of the Q-table is first initialized randomly and then learned via the following Q-function (or the Bellman equation):

$$\underbrace{\text{New}Q(s,a)}_{\text{New Q-Value}} = Q(s,a) + \underbrace{\alpha}_{\text{learning rate}} [\underbrace{R(s,a)}_{\text{Immediate reward}} + \underbrace{\gamma}_{\text{Discount rate}} \overbrace{\max Q'(s',a')}^{\substack{\text{Maximum predicted reward, given} \\ \text{new state s' and all its possible actions a'}}} - Q(s,a)]$$

where $R(s,a)$ is the immediate reward received when selecting the action a in the state s, α is the learning rate ($0 < \alpha \leq 1$) and γ is the discount rate ($0 < \gamma \leq 1$), reflecting the importance of the immediate reward comparing to the future rewards.

The reward function $R(s,a)$ is detailled in the Sect. 4.2.

3 Our Approach and Contributions

We propose an approach that learns how to assign learning activities to students in order to maximize their learning gains. This approach is based on connecting the Q-Learning to the BKT. Our contributions are:

1. We formulate the problem of sequencing of learning activities in order to maximize the student learning gain,
2. We propose a method based on simulated students to initialize the Q-table with "acceptable" values and answer the "cold start" problem of the Q-Learning algorithm.
3. We carried out a first experiment in order to evaluate the performance of the implemented system

4 Problem Formulation

In this section, we formalize the problem of learning a sequencing policy for maximizing the student learning gain. It implies to define the mains components of the Q-Learning algorithm: the states and the actions of the Q-table and the reward function that maximizes the student learning gain.

4.1 Student Knowledge State and Learning Activities

In accordance with the BKT model, each knowledge component (KC) is either in the learning state or in the unlearned state. Thus, we consider a knowledge state of a student as a vector of the mastering of each KC by the student (1 if the KC is learned by the student, 0 otherwise). The size of the vector is the number of the KC considered in the ITS. Thus, if we consider N KCs then we have 2^N possible knowledge states.

In Q-Learning, the Q-table is used to associate actions to states. In our case, we consider each student knowledge state as a state of the Q-table and each learning activity (definition, example, demonstration, etc.) as a possible action. Each time a learning activity is proposed to the student, an associated testing activity is also proposed. The testing activity is mandatory and the student has to perform it before passing to new learning activity. That is serves us for updating the mastery of the KCs in the BKT model of the student.

4.2 Reward Function

In each step for each student, the Q-Leaning selects a learning activity to present to the student, based on the Q-table and the ϵ-greedy exploration/exploitation strategy. Once, the student performs the testing activity associated to the selected learning activity, the BKT model infers the new mastery of the KCs worked on the learning activity. After converting the mastery probabilities to a binary values, the binary knowledge state is communicated to the Q-Learning agent for determining the next state in the Q-table. Finaly, the agent recieves the immediate reward corresponding to the move from the knowledge state s to the knowledge state s' and updates the Q table entry according to the Bellman equation (cf. Sect. 2).

The reward function is defined as the following:

$$R(s, a) = \sum_{i=1}^{N} (s'_i - s_i) \quad \text{if } s'_i > s_i$$

where s' is the new knowledge state of the student inferred by the BKT model after selecting the learning activity a and N the number of KCs in the ITS. The underlying idea of this reward function is that the more new KCs are mastered, the greater the reward. The cumulatives rewards quantify the learning gains of the students.

5 3-Step Process for Learning a Sequencing Policy

In education, it is quite critical to initialize the Q-table randomly because the RL agent, before learning enough a good sequencing policy, can recommand activities that are not well adapted to the students who may thus have to complete more activities than later students, may spend more time to improve their knowledge state and may be demotivated to use the ITS. In order to adress this concern-known as the "cold start" problem- we initialize the sequencing policy using simulated students. Indeed, several research have demonstrated that learning of the RL policy can be speeded up if the Q-table is initialized with relevant values. The aim is to have a balance between exploration and exploitation in the Q-Learning process.

Thus, we have implemented a 3-step process:

1. An initializing step: it consists on initializing the Q-table with simulated students. Rules were defined with experts to generate simulated students (cf. Sect. 6.1). These simulated data were used to train for the first time the sequencing policy and initialize the Q-table. In this step, the Q-Learning starts out by exploring random learning activities and exploits less.
2. A training step: in this step, the RL agent interacts with real students either by exploiting optimal decisions or by exploring other activities and updating the Q-table based on the expectation of the future rewards. These successive updates would allow the RL agent to converge to a good sequencing policy that maximize efficiently the students' learning gains. In this step, the Q-Learning exploits more and explores less.
3. A using step: when the RL agent has converged to a good sequencing strategy, it is time to use it to teach other students. These students will achieve their knowledge goals in the best way the Q-Learning agent has learned. In this step, the Q-Learning exploits most of the time because we have confidence in the learned sequency strategy.

6 Experimental Study

We carried out a first experiment with fourty students who interacted with the system, all of them are high school students starting to learn programming

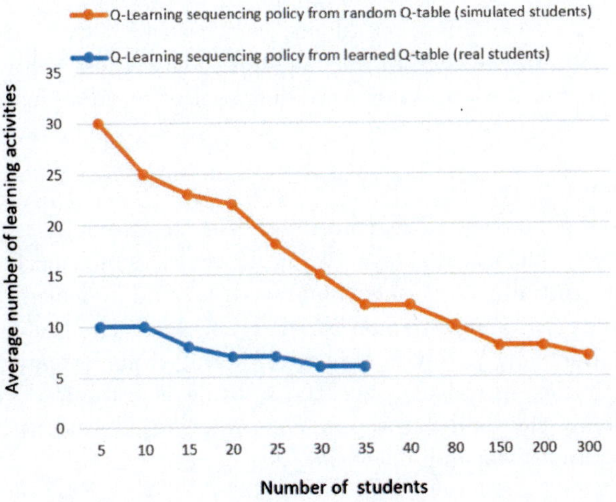

Fig. 1. Average number of learning activities displayed to the students in the case of sequencing policy learned from the Q-table initialized with simulated students and in the case of the sequencing policy learned from the Q-table initialized randomly.

with python. The knowledge domain is composed of the following KCs: "variable","sequential execution", "conditional structure"and "repetition structure". The objective of this first study is to answer the following research question: Is the initialization of the Q-table with data of simulated students will allow real students to achieve efficiently the mastering of all the KC in comparaison with a Q-table randomly initialized?

For answering this question, we compared the convergence of two sequencing policies: the first policy learned with simulated students from a random-initialized Q-table and the second one learned with real data from the Q-table obtained after the first policy.

6.1 Simulated Students

We model three student classes (strong, medium or weak) based on information provided by human experts about the number of attempts to answer correctly a testing activity and the prerequisite links between KC worked on in the ITS. Two rules were defined:

- R1: a student cannot answer correctly a testing activity on a KC without having first mastered all its prerequisite KC.
- R2: the probability that a student in the strong class answers correctly a testing activity is much higher than that of a student in the medium class and that of the latter is much higher than that of a student in the weak class.

These two rules were used to generate simulated data (500 simulated students in each class). They determine with fixed probabilities the correctness of a the student answer. These simulated students were used to learn a first sequencing policy in order to initialize the Q-table. We notice that these rules are only used to generate simulated data and not to learn to select learning activities.

6.2 Results

Figure 1 shows the number of learning activities required by the students to master the content. The x axis shows the number of students that have interacted with the system. Initially, the Q-Learning needs around 30 learning activities to allow the simulated students to master all KCs. After the first 300 simulated students, the sequencing policy is tuned, obtaining a performance of less than 10 learning activities. However, once the Q-table is initialized thanks to the simulated students, the sequencing policy allows the real students to master all KCs with less than 10 learning activities.

7 Conclusion

Adaptive sequencing of learning activities is crucial for improving students' learning gains in ITS.

This paper establishes connections between the Q-learning and the BKT and show that this mixed apporoach provides a potential solution. The 3-step method we propose could promote the use RL approaches in education.

The obtained preliminary results need to be further tested with other experiments by controlling variables such as the initial level of the students and even by using other individualized BKT models. There are several research directions for future work to provide evidence about the scaling-up of the approach.

References

1. Aleven, V., et al.: Instruction based on adaptive learning technologies. Handbook of Research on Learning and Instruction, pp. 522–560 (2016)
2. Bassen, J., et al.: Reinforcement learning for the adaptive scheduling of educational activities. In: Proceedings of the 2020 CHI Conference on Human Factors in Computing Systems (2020)
3. Corbett, A.T., Anderson, J.R.: Knowledge tracing: modeling the acquisition of procedural knowledge. User Model. User-Adap. Inter. **4**(4), 253–278 (1994). https://doi.org/10.1007/BF01099821
4. Doroudi, S., et al.: Sequence matters but how exactly? a method for evaluating activity sequences from data. In: Grantee Submission (2016)
5. Doroudi, S., Aleven, V., Brunskill, E.: Where's the reward? Int. J. Artif. Intell. Educ. **29**(4), 568–620 (2019). https://doi.org/10.1007/s40593-019-00187-x
6. Efremov, A., Ghosh, A., Singla, A.: Zero-shot learning of hint policy via reinforcement learning and program synthesis. In: International Educational Data Mining Society (2020)
7. Mandel, T., et al.: Offline policy evaluation across representations with applications to educational games. In: AAMAS, vol. 1077 (2014)
8. Watkins, C.J.C.H.: Learning from delayed rewards (1989)

Effects of VR on Learning Experience and Success

Stella Kolarik, Katharina Ziolkowski(✉), and Christoph Schlüter

Fraunhofer IML, Joseph-von-Fraunhofer-Straße 2-4, 44227 Dortmund, Germany
{Stella.Kolarik,Katharina.Ziolkowski, Christoph.Schlueter}@iml.fraunhofer.de
https://www.iml.fraunhofer.de/

Abstract. Virtual Reality (VR) and Serious Games have proven to be effective in employee training, especially when teaching practice-oriented tasks. This is of great value for intralogistics and its practical and complex processes. The VR-based learning game *InGo* simulates a receiving goods process for intralogistics. A study carried out in this work explores the effects on the learning success as well as other factors when learning with the VR-based game. It compares the effects with a control group that learns the same process with traditional text-based learning. Although the results of the study show that there is no significant difference between the two groups in terms of learning success itself, they confirm the positive influence on other factors that reinforce learning, such as intrinsic motivation and flow. This supports previous research that indicates the high potential of VR-based learning and educational games. The different preconditions of the VR group compared to the control group emphasize the positive effect of VR.

Keywords: Human-computer interface · Virtual reality · Learning study · Vocational training · Serious games · Intralogistics · Interactive learning environments · Technology enhanced learning

1 Introduction

In recent years with the rise of virtual and augmented reality there has been a strong increase in their use in education, especially in educational serious games [19]. Serious games are designed for a goal other than pure entertainment, often being educational or instructional [20].

Serious gaming has shown positive effects on learning experiences and outcomes [2]. The higher the virtuality of a learning experience, the deeper the participants immerse in the simulated working environment and thus they are less distracted [9]. This leads to the assumption that a VR-based serious game might have further positive effects on the learning success, especially in a practice-dominated area like intralogistics processes.

The VR-based educational serious game *InGo* (**In**coming **Go**ods) demonstrates the intralogistics process of incoming goods. It has been examined in a

M. Temperini et al. (Eds.): MIS4TEL 2022, LNNS 580, pp. 103–112, 2023.
https://doi.org/10.1007/978-3-031-20617-7_14

previous study in terms of physical and cognitive ergonomics with positive findings regarding different aspects of the learning experience [20]. However it has to be noted that this study did not compare the merits of *InGo* to traditional training methods. This is the main goal of this work: Comparing the effect of *InGo* on learning experience and success to traditional (e.g. paper-based) learning.

1.1 Background and Research Questions

Constructivism, a commonly used learning theory, assumes that learning is the way we construct meaning from experiences [19]. The more realistic these experiences are, the more efficient this process becomes at constructing a meaningful model of reality. Based on that VR could be a facilitator for learning as it allows for more realistic experiences than for example paper-based learning.

There have been indications that VR can enhance student performance in comparison to other, traditional learning methods in past research [15]. Still VR requires time to get used to, which can delay learning effects [20]. Thus we want to find out how VR and paper-based learning methods differ concerning the learning success (see Research Question *R1* below).

The use of VR systems can significantly improve different aspects of participant mood in comparison to pen-and-paper activities and also affect certain types of learning [13]. Therefore we want to find out how these three factors are connected (*R2*).

In the context of learning it has been shown that prior domain knowledge can influence learning effects [6]. We want to find out how prior knowledge affects the relationship between learning method and learning success in our case(*R3*).

One of the drawbacks of VR can be simulator sickness. It has been shown to have a negative impact on performance [24], but not necessarily on learning success [7]. We want to test how the learning methods are related to simulator sickness and whether the learning success is influenced by simulator sickness(*R4*).

Past research indicates that virtual learning environments such as VR can significantly increase motivation [3] in comparison to paper- or video-based learning [18]. High intrinsic motivation is associated with better learning results in long-term studies [26]. We want to find out how VR learning, intrinsic motivation and learning success are related (*R5*).

According to Cognitive Load Theory, a commonly used framework in learning and instruction, there is a limited capacity to our mental resources at any given moment and the learning efficiency is strongly connected to the way we use these resources [8]. Since VR offers a richer sensory experience in comparison to other media, there also are more stimuli to process, which can lead to a higher cognitive load [1]. Furthermore having to learn how to use a VR device can add to the cognitive load in the learning situation [20]. VR with clear instructions might be able to lower the perceived task load in comparison to a basic text. We want to find out about the relationship between the cognitive load, learning method and learning success (*R6*).

Flow is a mental state that is experienced when skills and challenge are approximately equal [16]. Individuals who experience flow tend to continue the activity simply because they enjoy it [26]. VR has been shown to foster flow [17], which in turn has been connected to learning [26]. We want to find out more about how learning method, perceived flow and learning success are connected (*R7*). In summary, the research questions are:

R1: Is there a difference between the use of VR and paper-based learning regarding learning success?
R2: What is the relationship between the use of VR, user mood and learning success?
R3: Does prior knowledge on the topic moderate the relationship between the learning method and learning success?
R4: What is the relationship between learning method, simulator sickness and learning success?
R5: What is the relationship between learning method, intrinsic motivation and learning success?
R6: What is the relationship between learning method, task load index and learning success?
R7: What is the relationship between learning method, flow experience and learning success?

2 Present Study

The goal of this study was to gain insights on the differences between traditional paper-based and VR-enhanced learning methods. For this we designed a study based on a serious game about the logistics topic of handling incoming goods. We cooperated with a vocational school that offers different courses in logistics. Two of their classes formed our respective groups. Both groups went through a process consisting of an initial questionnaire, an intervention and a final questionnaire, all completed within a single session. The entire study was conducted in German. The classes were randomly assigned to the VR-group or control group.

2.1 Study Design

The initial questionnaire was structured as following: First we asked the students for their age and gender. Then we asked about simulator sickness on a four-point-scale [22] and measured positive and negative mood aspects with a shortened PANAS five-point-scale [10]. We also questioned the participants about their prior knowledge concerning VR and incoming goods processes. For that we adapted two questions from Shou and Olney [21], which measure familiarity and experience on five-point-scales.

The intervention followed, during which both groups had to learn about the same topic in different ways: The VR-group played *InGo* on Oculus Quest 2 VR-devices provided by the school and the control group received traditional learning materials (text and images taken from *InGo*) on paper. Participants were asked to repeat and revise the content as often as they needed before proceeding. Both

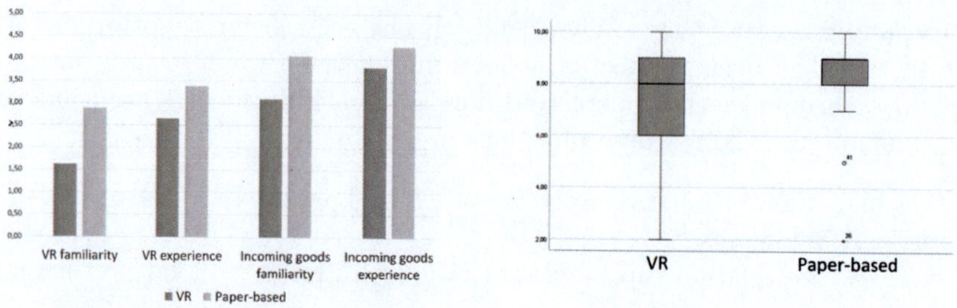

(a) Experience and familiarity.

(b) Knowledge test results.

Fig. 1. Values for previous domain knowledge and familiarity vs knowledge test outcomes for both groups.

the VR and control group were handled in supervised group settings of at least 10 students sitting or playing next to each other in parallel and instructed to fill out the questionnaires by themselves. The VR group received additional technical support from the research team in case of problems arising from the VR setup.

The second questionnaire followed after the intervention. Simulator sickness and mood were measured again. Next we measured intrinsic motivation with a translated KIM scale on a five-point-Likert-scale [23]. We measured the perceived task load with the translated NASA raw TLX [12]. The rTLX was measured on a scale from 0 to 100. The perceived flow was measured on a seven-point-Likert-scale [4]. Lastly a multiple choice test based on the knowledge to be acquired during the intervention was used to measure the learning success, similarly to past research [14]. For each correct answer they gained a point. Multiple marked answers, own additions or no answers were considered as wrong answers. The points were afterwards added together to form the result.

2.2 Results

Demographic Data. The VR group had 18 male and three female participants. The average age was $M = 21.19$, $SD = 2.75$. The control group had 17 male and three female participants. The average age was $M = 21.5$, $SD = 2.72$. There was no significant difference between the two groups concerning gender $t(39) = -.063$, $p = .950$ or age $t(39) = -.362$, $p = .719$.

However there were significant differences between the groups regarding VR experience $t(39) = -2.458$, $p = .019$, VR familiarity $t(39) = -3.138$, $p = .003$ and familiarity with incoming goods processes $t(39) = -2.781$, $p = .008$. Process experience showed no significant difference $t(39) = -1.711$, $p = .095$. The values for familiarity and experience of both groups can be seen in Fig. 1a.

Analysis of the Research Questions. Several participants left out single items or whole scales. If participants left out single items the whole scale was disregarded to avoid bias [11].

Table 1. Mean and standard deviation values for learning success and mood.

Group	Learning success		Positive mood before		Negative mood before		Positive mood after		Negative mood after	
	M	*SD*	*M*	*SD*	*M*	*SD*	*M*	*SD*	*M*	*SD*
VR	7.38	2.36	13.48	4.56	7.00	2.58	15.19	5.19	6.05	2.38
Paper-based	8.35	1.90	12.56	4.53	6.95	2.09	11.88	4.03	6.89	2.18

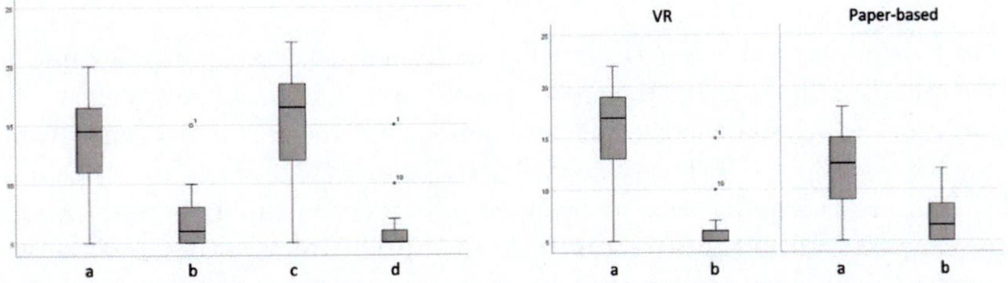

(a) Mood change by the VR intervention (a/b = +/- before, c/d = +/- after).

(b) Difference in mood between groups after the intervention (a = +, b = -).

Fig. 2. Boxplots showing the changes to positive (+) and negative (–) mood aspects before and after an intervention (left) and between groups (right).

R1. There was no significant difference between the two groups concerning the learning success meaning their test results, $t(39) = -1.446$, $p = .156$. The mean values and standard deviations of the test results among the values of the mood variables can be seen in Table 1. The paper-based group achieved slightly higher scores in the knowledge test than the VR group.

R2. Regarding the second research question, there were different aspects regarding mood, grouping and the learning success to be explored. As seen in Table 1 positive moods were significantly enhanced after the use of VR in comparison to the mood before, $t(20) = -4.683$, $p < .001$. Negative moods were significantly reduced after the use of VR $t(19) = 2.468$, $p = .011$ (Fig. 2a). There was a significant difference between the two groups concerning the positive mood after the task, $t(38) = 2.112$, $p = .042$, while there was no significant difference between the two groups concerning negative moods after the task, $t(38) = -1.170$, $p = .249$ (Fig. 2b). We then looked at the mediating influence of the mood regarding the two groups performance on the knowledge test. The analysis was conducted with the PROCESS macro by Hayes [5]. As seen in Fig. 3 there was no effect observed from the learning method on the learning success. Though newer works suggest this effect is not necessary for a mediation [25]. The two other paths were significant. Still we found no significant indirect effect, thus the relationship between learning method and learning success is not mediated by positive mood, indirect effect $ab = -.5954$, 95%-CI [−1.3178, .0206]. The relationship between learning method and learning success was also not mediated through negative mood as none of the paths were significant.

R3. For the third research question the moderation of the relationship between the learning method and the learning success through the prior knowl-

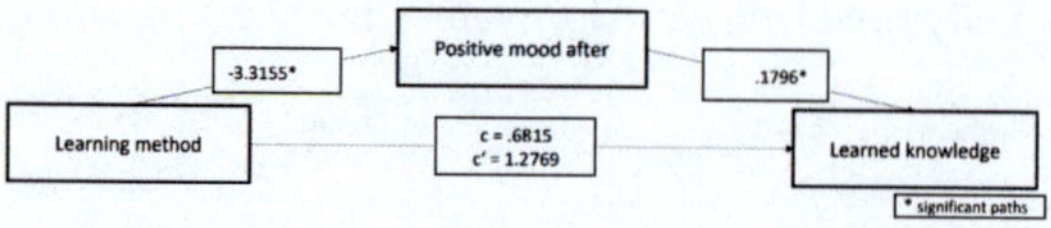

Fig. 3. The mediation of learning method, positive mood and learning success.

edge about incoming goods was tested. The analysis showed an almost significant overall model F(3,35) = 2.70, p = .060, predicting 39.49% of the variance. Yet prior incoming goods knowledge did not significantly moderate the relationship $\Delta R^2 = 0.3\%. F(1, 35) = .2026, p = .655$, 95%-CI [−.4348, .6751].

R4. This research question considered the relationship between learning method, simulator sickness after the task and the learning success. There was no significant difference between the two groups concerning simulator sickness, $t(35) = -1.234$, $p = .225$. The VR group experienced slightly less simulator sickness symptoms (Table 2, Fig. 4a). The relationship between learning method and learning success was not mediated by simulator sickness.

R5. Next we focused on intrinsic motivation in different learning situations. The VR group had significantly more intrinsic motivation for their task than the paper-based group, $t(31) = 2.254$, $p = .016$ (Table 2, Fig. 4b). We did not find a significant mediation of the relationship between learning method and success.

R6. There was no significant difference between the two groups in the perceived task load, $t(34) = -1.406$, $p = .169$. The VR group perceived a slightly lower task load (Table 2, Fig. 4c). The task load index did not mediate the relationship between the learning method and the learning success significantly.

R7. The last dimension we focused on was flow. As seen in Table 2 and Fig. 4d the VR group experienced significantly more flow than the paper-based group, $t(29.309) = 2.326$, $p = .014$. We inspected whether flow mediates the relationship between the learning method and the learning success. We did not find a significant mediation.

3 Discussion

The questions at the heart of this work were: *What impact does the use of VR have on learning success in comparison to traditional paper-based learning?* and *What are factors that influence this effect?* According to our results, VR had no direct significant effect on learning success. This indicates that VR and

Table 2. Mean and standard deviation values for simulator sickness, intrinsic motivation, task load and flow.

Group	Simulator sickness before		Simulator sickness after		Intrinsic motivation		Task load index		Flow	
	M	*SD*	*M*	*SD*	*M*	*SD*	*M*	*SD*	*M*	*SD*
VR	1.30	.30	1.26	.30	3.96	.46	34.42	13.02	4.65	2.18
Paper-based	1.41	.37	1.38	.31	3.49	.73	40.99	15.00	3.35	1.12

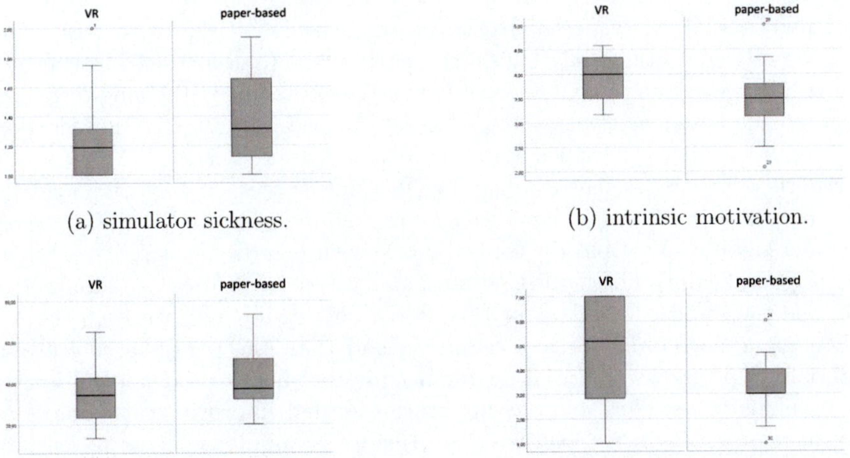

(a) simulator sickness.

(b) intrinsic motivation.

(c) perceived task load.

(d) perceived flow.

Fig. 4. Boxplots on differences between groups.

traditional paper-based learning methods were equally effective at conveying knowledge. It has to be noted though that the control group had a significantly higher prior familiarity with the topic than the VR group, which could have led to significantly higher results in the knowledge test in this group [6], but didn't. Therefore, the VR group might have performed better than the paper-based group if it had the same prior knowledge on the topic. Furthermore the VR group reached significantly positive results concerning moods, intrinsic motivation and flow, which are all facilitating factors in learning [26]. These findings are in accord with previous studies on the learning experience with VR [15,26], which found that while VR may not improve learning success in the short-term, it certainly can in the long-term.

Since the mood was measured both before and after each intervention, it was possible to compare within and between groups. The results show that there has been an improvement in both positive and negative mood aspects within the VR group from before to after the intervention. Also, the positive mood after the task was stronger in the VR group than in the paper-based group. In the context of learning this is an important finding as mood is known to be an important predictor in long-term learning outcomes [13].

There was no significant difference between the VR and the paper-based group concerning simulator sickness. In the context of learning, there is little evidence that would connect higher simulator sickness values to lower learning outcomes [7], even though it may affect the performance while using the simulator [24]. This indicates that simulator sickness might not be an obstacle in the adoption of VR for learning purposes.

We observed a significantly positive impact of VR on intrinsic motivation in comparison to the control group. However, we could not confirm a direct or mediating influence of this aspect onto the learning success of the two groups. This could be related to the relatively short length of the intervention. Motivational factors have been known to influence the learning outcome in the long-term [26]. Thus, an increased motivation could be interpreted as a positive influence on the overall learning experience with possible long-term effects.

Another important metric regarding learning success is cognitive load [8], which was measured using the perceived task load index. Since the VR group did not differ significantly from the control group with regards to cognitive load and the design and control of the VR serious game was more complex and unfamiliar than the paper-based method in this study, this could indicate that the VR serious game accounted for less cognitive load than the paper-based learning materials. This shows potential for further improvement of VR applications: if VR familiarity increases and the application design improves, this could lead to a decrease in cognitive load which in turn is strongly connected to learning outcomes.

Flow was one of the aspects that was significantly improved by the use of VR, but again did not predict the learning success. This does not confirm findings from other sources that suggest a strong interconnection between flow and motivational aspects which in turn have a strong positive influence on long-term learning success [26]. Similarly to the motivational aspect this might be explained by the short time of the intervention as well as the directly subsequent testing which allowed for assessment of short-term learning effects only. However it can be deduced that a significant increase in flow itself indicates an improvement of the learning experience.

3.1 Limitations and Future Research

A major limitation was conducting the studies in a group setting. A study design based on single participants would improve on this issue and enable us to diversify the groups to minimise differences regarding prior knowledge among others. Other test designs differing from paper-based learning (e.g. oral or practice exams) to cancel out imbalances due to structural similarity also would have been possible. However, this wasn't possible in the given organizational context at a public school with limited resources regarding staff, time and available rooms.

Moreover several participants did not fill out whole questionnaires and therefore reduced the amount of usable data. To prevent this in the future more participants could be recruited, additional approaches for promoting conscientiousness in participants could be implemented in the study design or an online questionnaire could be used that does not allow leaving questions blank as easily.

One major limitation of this study was that the knowledge test was to be filled out immediately after the intervention and thus uncovering only short-term learning effects. In order to asses the whole range of effects of VR-based learning however it's important to assess long term effects as well. Past research indicates that VR can be helpful in creating virtual "memory spaces" and thus facilitating memory recollection [19]. The study results state the improvement of moods, flow and intrinsic motivation by the use of VR, which have positive effects on long-term learning success [26]. Also as VR is rather uncommon in schools and takes time to get used to, different results could possibly be obtained in a longer study with repeated interventions [20].

The last important aspect is the cognitive load, which was measured using the task load index (NASA rTLX), which is limited to a specific point in time. We have only selectively summarized and evaluated information about a matter that is continuously changing in the course of the learning process. This also applies to other aspects like mood or flow. There are several options to measure these aspects continuously, e.g. with eye tracking, video based approaches or different physiological measurements like skin conductivity, heart monitoring or neuroimaging. This might help better understand the structural differences between VR and other methods in education when it comes to these aspects, which are, as our research suggests, key factors in learning success.

Acknowledgements. We would like to extend our thanks to the students and teachers at the *Berufskolleg Wirtschaft und Verwaltung Remscheid*. This study was carried out as a project within the interdisciplinary *Center of Excellence Logistics and IT* (https://leistungszentrum-logistik-it.de/).

References

1. Albus, P., Vogt, A., Seufert, T.: Signaling in virtual reality influences learning outcome and cognitive load. Comput. Educ. **166**, 104154 (2021)
2. Clark, D.B., Tanner-Smith, E.E., Killingsworth, S.S.: Digital games, design, and learning: a systematic review and meta-analysis. Rev. Educ. Res. **86**(1), 79–122 (2016)
3. Dalgarno, B., Lee, M.J.W.: What are the learning affordances of 3-d virtual environments? Br. J. Educ. Technol. **41**, 10–32 (2010)
4. Georgiou, Y., Kyza, E.A.: The development and validation of the ARI questionnaire: an instrument for measuring immersion in location-based augmented reality settings. Int. J. Hum. Comput. Stud. **98**, 24–37 (2017)
5. Hayes, A.F.: Introduction to mediation, moderation, and conditional process analysis: a regression-based approach. J. Educ. Meas. **51**(3), 335–337 (2014)
6. Johnson, A.M., Ozogul, G., Reisslein, M.: Supporting multimedia learning with visual signalling and animated pedagogical agent: moderating effects of prior knowledge. J. Comput. Assist. Learn. **31**, 97–115 (2015)
7. Johnson, D.M.: Introduction to and review of simulator sickness research (2005)
8. Korbach, A., Brünken, R., Park, B.: Measurement of cognitive load in multimedia learning: a comparison of different objective measures. Instr. Sci. **45**, 515–536 (2017). https://doi.org/10.1007/s11251-017-9413-5
9. Lamb, R.L.: Examination of the effects of dimensionality on cognitive processing in science: a computational modeling experiment comparing online laboratory simulations and serious educational games. J. Sci. Educ. Technol. **25**(1), 1–15 (2016). https://doi.org/10.1007/s10956-015-9587-z
10. Mackinnon, A., Jorm, A.F., Christensen, H., Korten, A.E., Jacomb, P.A., Rodgers, B.: A short form of the positive and negative affect schedule: evaluation of factorial validity and invariance across demographic variables in a community sample. Pers. Individ. Differ. **27**(3), 405–416 (1999)

11. Mazza, G.L., Enders, C.K., Ruehlman, L.S.: Addressing item-level missing data: a comparison of proration and full information maximum likelihood estimation. Multivar. Behav. Res. **50**(5), 504–519 (2015)
12. Moroney, W., Biers, D., Eggemeier, F., Mitchell, J.: A comparison of two scoring procedures with the nasa task load index in a simulated flight task. In: Proceedings of the IEEE 1992 National Aerospace and Electronics Conference@m_NAECON 1992, pp. 734–740, vol. 2 (1992)
13. Nadler, R.T., Rabi, R., Minda, J.P.: Better mood and better performance: learning rule-described categories is enhanced by positive mood. Psychol. Sci. **21**(12), 1770–1776 (2010)
14. Omelicheva, M.Y., Avdeyeva, O.: Teaching with lecture or debate? testing the effectiveness of traditional versus active learning methods of instruction. PS: Polit. Sci. Polit. **41**(3), 603–607 (2008)
15. Ray, A.B., Deb, S.: Smartphone based virtual reality systems in classroom teaching - a study on the effects of learning outcome. In: 2016 IEEE 8th International Conference on Technology for Education (T4E), pp. 68–71 (2016)
16. Reid, D.: A model of playfulness and flow in virtual reality interactions. Presence Teleoperators Virtual Environ. **13**(4), 451–462 (2004)
17. Ruvimova, A., Kim, J., Fritz, T., Hancock, M., Shepherd, D.C.: "Transport me away": fostering flow in open offices through virtual reality. In: Proceedings of the 2020 CHI Conference on Human Factors in Computing Systems, pp. 1–14 (2020)
18. Sattar, M.U., Palaniappan, S., Lokman, A., Hassan, A., Shah, N., Riaz, Z.: Effects of virtual reality training on medical students' learning motivation and competency. Pak. J. Med. Sci. **35**(3), 852 (2019)
19. Scavarelli, A., Arya, A., Teather, R.J.: Virtual reality and augmented reality in social learning spaces: a literature review. Virtual Reality **25**(1), 257–277 (2020). https://doi.org/10.1007/s10055-020-00444-8
20. Schlüter, C., Kretschmer, V.: Next level training in logistics: evaluation of a virtual reality-based serious game for warehouse logistics. Proceedings of the 19th International Conference on Modeling & Applied Simulation (MAS 2020), pp. 138–145 (2020)
21. Shou, Y., Olney, J.: Attitudes toward risk and uncertainty: the role of subjective knowledge and affect. J. Behav. Decis. Making **34**(3), 393–404 (2020)
22. Tauscher, J.P., et al.: Exploring neural and peripheral physiological correlates of simulator sickness. Comput. Animation Virtual Worlds **31**(4–5), e1953 (2020)
23. Wilde, M., Bätz, K., Kovaleva, A., Urhahne, D.: Überprüfung einer kurzskala intrinsischer motivation (kim) (2009)
24. Yörük Açıkel, B., Turhan, U., Akbulut, Y.: Effect of multitasking on simulator sickness and performance in 3d aerodrome control training. Simul. Gaming **49**(1), 27–49 (2018)
25. Zhao, X., Lynch, J.G., Chen, Q.: Reconsidering baron and kenny: myths and truths about mediation analysis. J. Consum. Res. **37**, 197–206 (2010)
26. Özhan, ŞÇ., Kocadere, S.A.: The effects of flow, emotional engagement, and motivation on success in a gamified online learning environment. J. Educ. Comput. Res. **57**(8), 2006–2031 (2020)

Educational Code-Review Tool: A First Glimpse

Zuzana Kubincová(✉), Ján Kl'uka, Martin Homola, and Adrián Marušák

Comenius University in Bratislava, Mlynská dolina, 842 48 Bratislava, Slovakia
{kubincova,kluka,homola}@fmph.uniba.sk, marusak2@uniba.sk

Abstract. Code review is a common part of a programmer's job and is considered a best practice when developing software projects. Recently, this technique has also found its place in educational activities. In addition to developing programming skills, it has been shown to bring other benefits to students, such as an increase in motivation, improvement of learning outcomes, development of soft skills, etc.

We have been using code review in our courses for several years. So far, however, we have been using a review tool that was not developed primarily for code review. In this paper, we describe a new version of our tool that has similar basic functionality to professional tools but is simpler and more suitable for use in education. Our first experiences from its pilot testing with students are also presented.

Keywords: Pedagogical code-review · Code-review tool · Education

1 Introduction

Programming courses are a highly important part of computer science curricula. Various methods are used to teach them, nevertheless programming remains a subject that many students find difficult and therefore they need to be well motivated to actively immerse in it.

For these reasons university curricula try novel methods and approaches to increase students' engagement and to improve the learning outcomes in programming courses. One of the methods that are gaining interest is *code review*. It has been shown to bring various benefits for students [11,12,14]: besides being an effective way of learning to code, it also helps to develop soft skills in students, which are nowadays required by many employers. In fact, code review – by itself – is an important skill that will become a regular part of their job as programmers in the IT industry. From our point of view, code review can serve as a motivating factor for students. Being a form of peer-review, it yields many of the benefits thereof [24–27]. By engaging students with their peers' code, it brings in a new and interesting activity, and encourages them to take a more active role in their education.

It is important to note that in education code review used must be conducted differently from how it is established in the IT industry: in the education

M. Temperini et al. (Eds.): MIS4TEL 2022, LNNS 580, pp. 113–122, 2023.
https://doi.org/10.1007/978-3-031-20617-7_15

setting the activities and the tools used must be properly designed to allow seamless integration in the educational process and to specifically reinforce the set learning goals. This is why various pedagogical code review methodologies are being developed [12,13,16–19], and educational professionals are designing and developing their own tools to use this technique in their courses.

A few years ago we started to incorporate code review into the teaching of programming. However, the tool we were using until then was developed to apply peer review on students work other than program code. In this paper, we present a new module that has been integrated with our LMS system and that has been specifically designed to execute pedagogical code review: The students are assigned a certain number of their peer's programming submissions for code review; the tool allows to display and browse the code and to add comments to any line or block of code. In comparison to professional code review environments (e.g., those integrated with GitHub and other platforms), the workflow is streamlined and anonymous reviewing is allowed to foster fairness. We describe the design of the tool and our first experience with it.

2 Related Work

2.1 Code Review

Code review is an essential part of the development of programming projects in software companies. It is an activity, in which a programmer reviews the code of a colleague's computer program in order to identify bugs that might have been overlooked, and possibly uncover inefficiencies in the program, point out how to incorporate best practices and coding standards, thus improving its quality [1,2]. This activity is usually done offline, whereby the reviewer reads the program code, tries to understand it and grasp its functionality without running the program. Code review, especially in large-scale software development projects, is performed simultaneously with program development because it is considerably cheaper to fix bugs during development than at later stages [3]. As shown in the Wiegers' study [4], code review can help to detect 50 to 70% of bugs.

The beginning of the use of code review in software development dates back to the 1970 s when the principles of "code inspection" were formulated [5], and subsequently, first references to this activity began to appear in the professional literature [6].

Since then, the procedures applied in code review have changed considerably. Many different methodologies are used, and new methodologies are continually being developed to improve both the process and the results of code review [3, 7–9]. Nowadays, code review has become less formal and less rigorous, but it also focuses on other aspects that may be beneficial to the software product development team, such as knowledge transfer, increased team awareness, and the creation of alternative solutions [10].

2.2 Pedagogical Code Review

In the context of the latter benefits of code review, in particular the transfer and acquisition of knowledge between the code author and the reviewer, educational practitioners and researchers have become interested in the use of code review in teaching and learning to code [11–13].

As many studies on the use of activities of this type in education have shown, code review can serve as an effective tool to improve the ability of students to write quality code, teach them coding standards in programming, and also provoke well-thought-out discussions of programming issues and practices [11,14]. In addition to the usefulness of code review for developing students' programming skills, the authors of several studies [12,15,16] report that this method of active learning does not only positively impact student learning in various aspects and improve student attitudes, but also serves to develop soft skills, such as communication skills, critical thinking, and team-work skills.

Similar to industrial code review, pedagogical code review uses a wide variety of methods [12,13,16–19]. These activities can be carried out by students individually or in teams, or both approaches can be used in successive rounds of code review [17]. Students may be strictly divided into a group of authors and a group of reviewers, or they may gradually be given the roles of author, reviewer, and reviser [16]. Pedagogical code review can be performed without the use of a tool, but it can also be tool-based. Since the tools used for professional code review usually do not meet the requirements for use in pedagogical code review [18], many researchers and practitioners in the educational environment develop their own tools [18,20–23].

3 Educational Setting

3.1 Courses

In the informatics teacher education program, programming is taught in the first three semesters of the bachelor's degree and is followed up with a course on Algorithms and Data Structures in the fourth semester. In each of the Programming 2 and Programming 3 courses, students are given a semester-long project in which they usually program a game. In doing so, they must apply their knowledge from the entire semester of programming. In the Algorithms and Data Structures course, students sequentially program three projects per semester, the goal of which is for them to better understand and practice specific data structures and the algorithms that operate on them.

3.2 Methodology

According to our previous experience with peer review in other courses, not all students were willing to participate in reviewing the work of their peers. To better motivate them to engage in this activity in the case of code review, the process of developing and reviewing programming projects was conducted in three steps:

a) programming the first version of the program,
b) reviewing a colleague's program,
c) improving the program.

After the project topic was assigned, students programmed their projects individually. They had several weeks to program a first version of the project, which they then submitted for code review. Each student in the Programming course who submitted their project in this first phase was assigned one of their classmates' projects to review. In the Algorithms and Data Structures course, we considered the students to be more proficient reviewers since they had already completed code review in previous courses, and therefore they always reviewed two of their classmates' projects in this course. A student who did not submit their project for review was not involved in the review process as a reviewer.

Students performed code review based on criteria prepared by the teacher in the form of questions supplemented with hints to better clarify which aspects of the program to focus on, e.g., clarity of the program code, correct handling of inputs, efficiency of the program, etc. Students rated each aspect of the program on a scale of 1–5 and also provided verbal comments to justify their ratings.

The students had approximately one week for code review, after which they submitted their reviews and these were delivered to the authors of the programs. The code review process was double blind, so neither the reviewer knew whose program they were reviewing, nor did the author know who was reviewing their program.

In the next phase of the project, the programmers had approximately one week to incorporate the reviewers' comments into their program. Only after this modification did they submit this improved version to the teacher for evaluation.

4 Previous Experience with a Peer-Review Tool

For the purpose of peer review, we have already developed a tool that has been integrated as a module into our own LMS courses.matfyz.sk. This module, called Assignments, allows the teacher to manage assignments, student projects and reviews, as well as to comment and grade them. It can also be used to create and manage student teams in case it is a team assignment. With this module, students can submit projects, review other students' projects, and also track the reviews they have received from their colleagues, as well as comments and assessments from the teacher.

Once an assignment is posted for students, the teacher will specify in the system whether it is an individual or team project, how many rounds the students will work on the project, and whether they will also submit an improved version after the reviews. Then, for each round, they set a deadline for submitting the first version of the project, a deadline for submitting the reviews, and a deadline for submitting the improved project. If it is a team project, the system also allows to do a team review – a peer review of each team member's work.

For reviewing, the teacher can set how many projects each student should review, whether it will be a blind or double-blind review, the form of the review (points, comments, or both), and specifies the questions and hints by which the students should rate the assigned projects.

Students are shown a timeline in the LMS with the scheduled deadlines. After the deadline for the first project submission has passed, the system randomly distributes the submissions among those students who have submitted a project and displays the review form to them.

During peer review process, project authors do not see the reviewers' comments and ratings. These will only be displayed to them after the review deadline has passed.

Even if reviewing is set as blind or double-blind, this functionality only limits students. The teacher can see who is reviewing whose project and can also see all submissions and reviews and respond to them (both submission and review) with a comment.

5 New Code-Review Tool

Since this tool was originally developed before we introduced code review into our courses, it does not have some specific features that would be useful for reviewing program code. Therefore, our goal was to develop a new version of this tool with enhanced functionality.

5.1 Commenting on Code Blocks

With the peer review tool used so far, students could only comment on a peer's program through a form in which they answered questions prepared by the teacher. If they wanted to express an opinion on a specific block of the program, e.g. a function or a sequence of program steps, they had to describe it properly in the comment so that it was obvious to the author what they were commenting on, and only then could they add their comment on that part of the code. This procedure would be greatly simplified if students could add comments directly to the program code. We have therefore added this functionality in the new version of our tool.

On the other hand, reviewing code based on the teacher's questions and hints is especially useful for students who have not yet performed such an activity or are not yet proficient in it, and thus need guidance from the teacher on what aspects to pay attention to when reviewing code. Therefore, the new version of the tool allows to use both methods of commenting code – forms, as well as adding comments directly to blocks of program code.

After the deadline for submitting the first version of the program, the student will see a page with the programs assigned to them for review. Clicking on any assigned program will display multiple tabs for this program. One of them shows the review form with questions from the teacher and another one shows the entire program code. The reviewer can mark a section of code and add a comment to it. This part can be as little as one line or as long as a block of commands consisting of several lines. The part of the program to which the comment is added is highlighted in color.

The tool also allows nested comments, i.e. a comment can be added to some part of a block which has already been comment on, and even this part of the block can contain other nested comments (up to the depth of 5). The nested comments are differently colored so that they are easily identifiable (Fig. 1).

Reviews, both in the form of answers to questions on the form and comments on code blocks, are displayed as anonymous to the author of the program if blind or double-blind reviewing has been selected by the teacher for the assignment. Reviews are always non-anonymous to the teacher.

The program's author can reply to comments left by the reviewer on the program blocks, the reviewer can reply to those replies, etc. (Fig. 1). Such a discussion of the code is still anonymous and can go on for as long as desired even after the review phase is over.

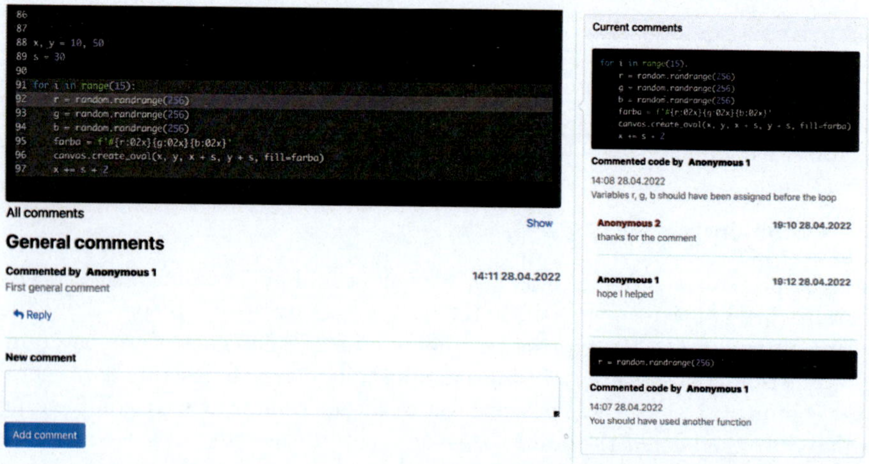

Fig. 1. Highlighted nested comments on a program code blocks (top left); General comment (bottom left); Reviewer and author discussion via comments (right).

5.2 Other New Features

General Comments. In addition to evaluating individual aspects of the program using the form and adding comments directly to the program code blocks, the student can also add a general comment (Fig. 1) to the entire program, in which they can express their overall opinion about the program code or write comments that are not related to specific blocks of the program.

General comments can still be added after the review phase is over, and allow the reviewer to leave a note to the program author even later. These late comments are also shown as anonymous to the author, who can reply to them. An anonymous discussion between the author and the reviewer is thus possible, just as in the case of comments on program blocks.

Showing Both Versions of the Program. Since it can be useful for the teacher to see both versions of a student's program, new tabs have been added to the teacher's view where both the original version and the final version can be displayed. The teacher can also see the comments that the reviewers added directly to the blocks of the initially submitted code and check how the student incorporated these comments into the improved version of the program. The teacher can also comment on both versions.

Review Indicator. Indicators of whether a program has already been reviewed or not have been added in order to improve the user experience. On the page where the student-reviewer sees the assigned programs, a cross indicates a program that has not yet been reviewed, while a check mark indicates an already reviewed program (see Fig. 2). Similarly, one of these two indicators is displayed to the author of the program after the review deadline has passed – depending on whether or not their program has been reviewed by that deadline (see Fig. 3).

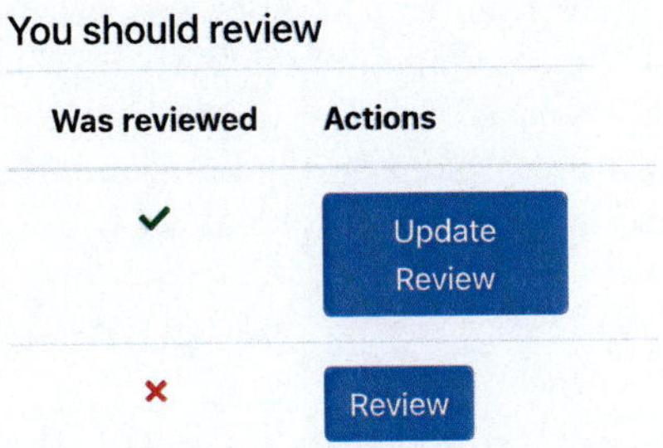

Fig. 2. Reviewer's view of a peer-reviewed and an unreviewed program.

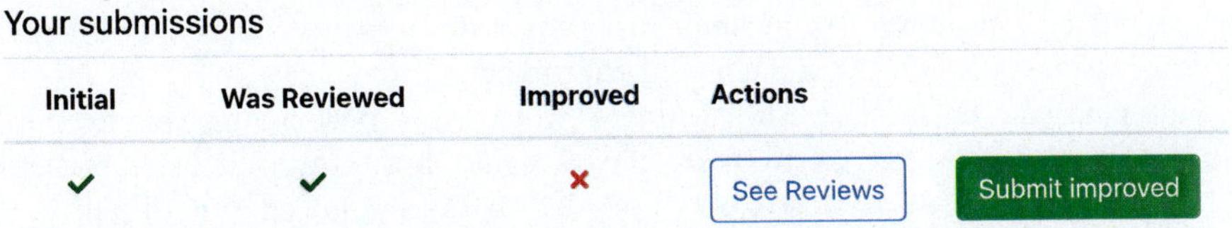

Fig. 3. Author's view of the state of their submission.

Review Update. For a program that has already been reviewed, the student-reviewer will see an "Update Review" button next to the check mark indicator, which allows them to edit the saved review (see Fig. 2) if the review deadline has not passed yet. After the review phase deadline, the reviewer will see a "View" button instead of "Review" or "Update Review" buttons. They can then view the reviews they have assigned to the program, but they can no longer change their review form responses. They can, however, add comments on code blocks, general comments, and replies to both kinds of comments, as mentioned above.

5.3 Testing

Pilot testing of the tool was performed in the Algorithms and Data Structures course with students who had previously used the original Assignments module.

The aim of testing was to focus not only on the usefulness of the module in terms of writing better and more specific code reviews, but also on its proper functioning and user experience. Therefore, the new functionality of the module was not explained to the students and they were left to navigate the new environment on their own.

After testing, students were asked to complete a short questionnaire. From the collected data, we discovered a number of shortcomings that made navigating the system a bit more difficult. Some of them have already been solved and some remain as future work.

Regarding the students' opinion on the usefulness of commenting code blocks, the students who discovered this possibility appreciated it as very helpful. They justified it as follows: "When using the old system, I occasionally wanted to comment on some part of the code, but it was too complicated to explain so that the author would understand which part I was writing about. That's why I sometimes gave up. So I think commenting directly on a block of code is a great idea. ... And I also think it's easier and less time-consuming for the author of the code. And so we can both focus more on the programming side of things rather than some laborious description of the problem area."

6 Conclusions

After several years of incorporating code review into programming courses, we designed and implemented an enhanced code-review tool integrated with our own LMS. In addition to the ability to evaluate aspects of program code by scoring and answering questions on a form, the new module also offers direct assignment of comments to blocks of program code, writing general comments, and discussion between the author and the reviewer.

Our goal was to simplify the students' work while performing this activity and thus motivate them to write more accurate and more comprehensive reviews of their peers' programs. Based on pilot testing the enhanced functionality of the tool with students who had used our previous peer review system before, we believe that the new tool accomplishes that goal. Students find the direct commenting on code blocks to be a great idea that will make it easier for them to review and for program authors to identify problems in their code that need to be fixed. They also expressed that it should become the main way of reviewing code in our system.

A limitation of this work is that the new tool has so far only been tested on a pilot basis. We consider its further testing and refinement an important part of our future work, as we plan to continue using the code review technique in our courses and want to motivate students to actively participate in this activity. We prove that this is of great importance to them by the words of our student: "When reviewing the code of others, I may notice how they program, whether less efficiently than I do or more. If less, I am happy to advise them, if more I am inspired by their coding. It is a win-win situation."

Acknowledgements. This research was supported by Slovak national projects VEGA 1/0621/22 and APVV-20-0353.

References

1. Fagan, M.: Design and code inspections to reduce errors in program development. In: Broy, M., Denert, E. (eds.) Software Pioneers, pp. 575–607. Springer, Heidelberg (2002). https://doi.org/10.1007/978-3-642-59412-0_35
2. Boehm, B.W., et al.: Software Engineering Economics. Prentice-Hall Englewood Cliffs, Hoboken, vol. 197 (1981)
3. Uwano, H., Nakamura, M., Monden, A., Matsumoto, K.I.: Analyzing individual performance of source code review using reviewers' eye movement. In: Proceedings of the 2006 Symposium on Eye Tracking Research & Applications, pp. 133–140. ACM (2006)
4. Wiegers, K.E.: Peer Reviews in Software: A Practical Guide. Addison-Wesley, Boston (2002)
5. Sadowski, C., et al.: Modern code review: a case study at google. In: Proceedings of the 40th International Conference on Software Engineering: Software Engineering in Practice, pp. 181–190. ACM (2018)
6. Fagan, M.E.: Advances in software inspections. In: Pioneers and Their Contributions to Software Engineering, p. 335 (1986)
7. Höst, M., Johansson, C.: Evaluation of code review methods through interviews and experimentation. J. Syst. Softw. **52**(2), 113–120 (2000)
8. Beller, M., et al.: Modern code reviews in open-source projects: which problems do they fix? In: Proceedings of the 11th Working Conference on Mining Software Repositories, pp. 202–211. ACM (2014)
9. Allen, H., Cury, A., Gaston, T., Graf, C., Wakley, H., Willis, M.: What does better peer review look like? Underlying principles and recommendations for better practice. Learned Publishing **32**(2), 163–175 (2019)
10. Bacchelli, A., Bird, C.: Expectations, outcomes, and challenges of modern code review. In: Proceedings of the 2013 International Conference on Software Engineering, pp. 712–721. IEEE Press (2013)
11. Hundhausen, C., Agrawal, A., Fairbrother, D., Trevisan, M.: Integrating pedagogical code reviews into a CS 1 course: an empirical study. In: ACM SIGCSE Bulletin, vol. 41, no. 1, pp. 291–295 (2009)
12. Hundhausen, C.D., Agrawal, A., Agarwal, P.: Talking about code: integrating pedagogical code reviews into early computing courses. ACM Trans. Comput. Educ. (TOCE) **13**(3), 1–28 (2013)
13. Indriasari, T.D., Luxton-Reilly, A., Denny, P.: A review of peer code review in higher education. ACM Trans. Comput. Educ. (TOCE) **20**(3), 1–25 (2020)
14. Li, X., Prasad, C.: Effectively teaching coding standards in programming. In: Proceedings of the 6th Conference on Information Technology Education, pp. 239–244. ACM (2005)
15. Grissom, S., Hundhausen, C., Conrad, P.: Alternatives to lecture: experience peer instruction and pedagogical code reviews. In: Proceedings of the 45th ACM Technical Symposium on Computer Science Education, pp. 275–276 (2014)

16. Wang, Y., Li, H., Feng, Y., Jiang, Y., Liu, Y.: Assessment of programming language learning based on peer code review model: implementation and experience report. Comput. Educ. **59**(2), 412–422 (2012)
17. Trytten, D.A.: A design for team peer code review. In: ACM SIGCSE Bulletin. ACM. pp. 455–459 (2005)
18. Kubincová, Z., Homola, M.: code review in computer science courses: take one. In: Xie, H., Popescu, E., Hancke, G., Fernández Manjón, B. (eds.) ICWL 2017. LNCS, vol. 10473, pp. 125–135. Springer, Cham (2017). https://doi.org/10.1007/978-3-319-66733-1_14
19. Almeida, F.: Framework for software code reviews and inspections in a classroom environment. Int. J. Mod. Educ. Comput. Sci. **11**(10), 31 (2018)
20. Hundhausen, C., Agrawal, A., Ryan, K.: The design of an online environment to support pedagogical code reviews. In Proceedings of the 41st ACM Technical Symposium on Computer Science Education, pp. 182–186 (2010)
21. Tang, M.: Caesar: a social code review tool for programming education. PhD thesis. Massachusetts Institute of Technology (2011)
22. Combéfis, S.: Automated code assessment for education: review, classification and perspectives on techniques and tools. Software **1**(1), 3–30 (2022)
23. Indriasari, T.D., Luxton-Reilly, A., Denny, P.: Improving student peer code review using gamification. In: Australasian Computing Education Conference, pp. 80–87 (2021)
24. Mulder, R.A., Pearce, J.M., Baik, C.: Peer review in higher education: student perceptions before and after participation. Act. Learn. High. Educ. **15**(2), 157–171 (2014)
25. Kubincová, Z., Homola, M., Bejdová, V.: Motivational effect of peer review in blog-based activities. In: Wang, J.-F., Lau, R. (eds.) ICWL 2013. LNCS, vol. 8167, pp. 194–203. Springer, Heidelberg (2013). https://doi.org/10.1007/978-3-642-41175-5_20
26. Lundstrom, K., Baker, W.: To give is better than to receive: the benefits of peer review to the reviewer's own writing. J. Second. Lang. Writ. **18**(1), 30–43 (2009)
27. Gielen, S., Tops, L., Dochy, F., et al.: A comparative study of peer and teacher feedback and of various peer feedback forms in a secondary school writing curriculum. Br. Edu. Res. J. **36**(1), 143–62 (2010)

Retrieving Key Topical Sentences with Topic-Aware BERT When Conducting Automated Essay Scoring

Yongchao Wu(✉), Aron Henriksson, Jalal Nouri, Martin Duneld, and Xiu Li

Stockholm University, NOD-huset, Borgarfjordsgatan 12, 16455 Stockholm, Sweden
{yongchao.wu,aronhen,jalal,xmartin,xiu.li}@dsv.su.se

Abstract. Automated Essay Scoring (AES) automatically assign scores to essays at scale and may help to support teachers' grading activities. Recently, AES methods based on deep neural networks (DNN) have significantly improved upon the state-of-the-art performance by learning relations between holistic essay scores and student essays. However, DNN-based AES methods function like black-box, negatively affecting the ability to provide automated writing evaluation (AWE). In this work, we proposed a new method, topic-aware BERT, based on fine-tuning the pre-trained language model to learn relations between essay scores and text representations of student essays as well as topical information in essay writing instructions. Moreover, we propose an approach to automatically retrieve key topical sentences in student essays by probing self-attention maps in intermediate layers of topic-aware BERT. We evaluate the performance of topic-aware BERT to (i) perform AES and (ii) retrieve key topical sentences using the open dataset Automated Student Assessment Prize and a manually annotated dataset, respectively. Our model achieves a strong AES performance compared with previous state-of-the-art DNN-based methods and shows effectiveness in identifying key topical sentences in argumentative essays.

Keywords: Natural language processing · Automated essay scoring · Automated writing evaluation · BERT

1 Introduction

Automated essay scoring (AES), which can support essay grading at scale, plays an essential role in lightening teachers' workload [1,2] and reducing intra-rater and inter-rater inconsistency [3,4]. Researchers have mainly investigated feature-based and deep neural network (DNN)-based approaches to AES. Expert knowledge is usually required for feature-based approaches to design linguistic indices or rubric-based features [5–7] fed into a regression machine learning model. In another line of research, various nonlinear neural networks have been used to automatically learn the relation between essay representations and their scores

M. Temperini et al. (Eds.): MIS4TEL 2022, LNNS 580, pp. 123–132, 2023.
https://doi.org/10.1007/978-3-031-20617-7_16

in an end-to-end fashion. For instance, Taghipour et al. [8] introduce AES methods based on Convolutional Neural Network and Long Short-Term Memory networks (CNN-LSTM) without manual feature engineering. Dong et al. [9] improve AES performance by proposing the use of global attention and hierarchical networks (CNN-LSTM-Att). Recently, large pre-trained language models, such as BERT [10], have shown remarkable performance in different natural language processing (NLP) tasks, including AES [11]. Despite the impressive performance of DNN-based AES systems, they work like a black box [5], negatively affecting the interpretability of the model. Due to the lack of interpretability, DNN-based approaches encounter difficulties when constructing automated writing evaluation (AWE) systems. AWE systems can assist teachers' grading and students' revision by providing enlightening feedback, and this would be easier to implement with feature-based approaches [12,13]. Zhang and Litman [14] address this problem by exploring the use of attention weights in their co-attention-LSTM-CNN architecture, aiming at providing AWE feedback on topical words and phrases level for a particular essay writing task response-to-text assessment (RTA). We note that no similar investigations have been conducted with BERT-based AES systems, while Clark et al. [15] have proved that self-attention maps in BERT could help understand what neural networks learn about language. From a pedagogical point of view, student academic success requires the ability to produce a high-quality argument, complete with statements, warrants, and evidence [16–18]. Together with a predicted score, we believe that spotting key topical sentences in argumentative essays could facilitate teachers' grading process when judging the quality of essays. In this spirit, this study aims to provide AWE feedback on topical sentences from argumentative essays by probing the self-attention mechanism in a BERT-based AES system. As defined by Newell et al. [19], argumentative writing refers to *writing in a principled way to support a claim using reasons and evidence from multiple sources.* In a practical scenario, students usually write argumentative essays by paying attention to topics in essay instructions. Some of the sentences in the essays reflect topical information either by reasons or evidence, and we define these sentences in this work as key topical sentences (KTS). Specifically for our method, we feed both topical information from the essay instruction and student essays to BERT to train AES system (denoted topic-aware BERT). Then we retrieve KTS by ranking the self-attention weights between sentences in student essays and topical information. We hypothesize that sentences with higher self-attention weights with keywords affect the model's grading prediction and thus play a similar role to KTS. To evaluate our method, we deploy an open dataset to evaluate the AES performance using the official metrics Quadratic Weighted Kappa (QWK). Besides, we manually create a dataset to evaluate the performance of the model to retrieve KTS. The results show that topic-aware BERT achieves a competitive AES performance compared to state-of-the-art models and that our KTS retrieving method is effective when applied to argumentative essays.

This paper is organised as follows. Section 2 presents the data used in this paper, including the ASAP dataset, topical keywords and a human-annotated

KTS dataset. We describe the method formulation of the proposed AES model and KTS retrieving approach in Sect. 3. Experiments settings, evaluation metrics and experiment results regarding AES as well as KTS performance are presented in Sect. 4. Finally, Sect. 5 concludes this paper and discusses future works.

2 Data

Table 1. ASAP dataset statistics.

Prompt	Essay size	Genre	Avg Len
1	1783	ARG	350
2	1800	ARG	350
3	1726	RTA	150
4	1772	RTA	150
5	1805	RTA	150
6	1800	RTA	150
7	1569	NAR	250
8	723	NAR	650

Table 2. Topical keywords extracted for each essay prompt from essay instructions. The essay instructions can be found in the original ASAP dataset descriptions.

Prompt	Topical keywords
1	Computer, positive, concern
2	Censorship, library
7	Patience, story

2.1 Automated Student Assessment Prize (ASAP) Dataset

We deploy the open dataset Automated Student Assessment Prize (ASAP) dataset[1] from the Kaggle competition to evaluate the automated essay scoring performance of our proposed approach. In this dataset, there are eight essay sets. Students have written essays according to independent prompts in different essay sets. The essays belong to different genres. Essays from prompts 1 and 2 are argumentative (ARG) essays; essays from prompt 3 to 6 are response-to-text (RTA); essays from prompts 7 and 8 are narrative (Narrative) essays. According to [5], only essay prompts 1, 2, 7, and 8 truly examined students' writing skills. Considering that we focus on argumentative essays, this study experimented with essays from prompts 1, 2, and 7. We include narrative essay prompt 7 because we would like to see how our KTS-retrieving method performs differently on argumentative and narrative essays. Prompt 8 essays are excluded in the experiment because they are too long, and some of the key topical sentences might be truncated by BERT. Some other ASAP dataset statistics can be found in Table 1.

2.2 Topical Keywords

We also extracted topical keywords from prompts 1, 2, and 7, which are shown in Table 2. Due to the limitation that BERT can only process up to 512 tokens [10],

[1] https://www.kaggle.com/c/asap-aes/data.

the topical keywords extracted from the essay instructions serves as the topical information. Three experts were recruited to pick the topical keywords from essay prompt 1, 2, and 7. Each expert was asked to pick up the topical keywords candidates which in their mind can cover most of the topical information from the essay prompt for each essay prompt. After that, we deploy the NLTK[2] toolkit to get the lemmatized versions of the topical keyword candidates. Then we take the intersection of the topical keyword candidates from each expert for each prompt, shown in Table 2.

2.3 Human Annotated KTS Dataset

To evaluate our KTS retrieving methods, we manually created an evaluation dataset through a clickable web-based annotation tool[3]. Firstly we recruited three experts to conduct a prestudy to define the annotation gold standard. During the prestudy, the experts were given 12 randomly selected essays (3 essays from each prompt) and asked to pick five KTS for each essay. From the result of the prestudy, we find that even though some annotations are different, they are related to the different aspects of the same topics. Based on this observation, we decide to take the union of two annotators' annotations as the gold topical key sentences for each essay. We employed six PhD students to annotate the same 12 essays following the same procedure to verify this idea and avoid motivation and knowledge bias. We calculate the union scores between the annotations from experts and PhDs. The average union score of annotations among experts and PhDs are 7 and 7.7, respectively, which means the range of average union score should be between 7.7 and 14.7. Thus the average union scores of annotations between expert and PhD groups (=9) is very close to the minimum score (=7.7), which indicates a high overlapping of annotations. This results confirm that it is reliable to use the union of two annotators' annotations to serve as gold KTS. Based on the recommendations that the minimum sufficient test size to evaluate an information retrieval system should be 50 [21], we recruited six PhD students to annotate 60 essays (20 essays from each prompt) randomly selected and in line with the original essay grade distributions following the guidelines from prestudy to evaluate our KTS retrieving method.

3 Methods

The overview of our proposed grading assistant model, topic-aware BERT, is shown in Fig. 1. By taking topical information from the essay prompts and student essays, topic-aware BERT can provide two kinds of information to assist teachers' grading: a predicted essay score indicating the general quality of the essay as well as the spotted key topical sentences of the essay.

---

[2] https://www.nltk.org/.

[3] The evaluation dataset and annotation tool source code could be provided as requisition via email.

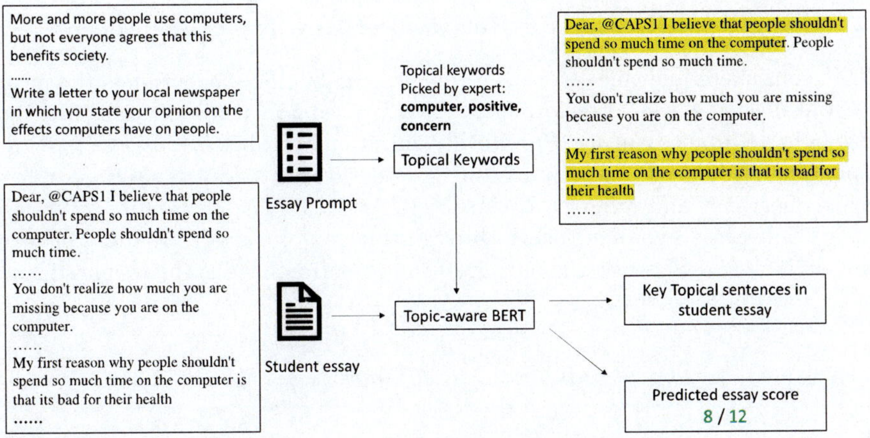

Fig. 1. Overview of the grading assistant framework (topic-aware BERT) for teachers. We trained a BERT-based AES model with both student essay and essay instruction as input, which can automatically predict essay scores and spot key topical sentences.

3.1 Topic-Aware BERT

To make BERT aware of topical information, we concatenate topical keyword tokens K $(t_{k_1}, t_{k_2}, \ldots, t_{k_n})$ and the whole essay tokens E $(t_{e_1}, t_{e_2}, \ldots, t_{e_M})$ with a special $[SEP]$ token to represent input sequence for BERT, denoted $S = [CLS], t_{k_1}, t_{k_2}, \ldots, t_{k_n}, [SEP], t_{e_1}, t_{e_2}, \ldots, t_{e_M}$, where n and M are the number of tokens in topical keywords and essay respectively. We feed the input sequence representation S into BERT and use the final hidden vector $h_{[CLS]} = C \in \mathbb{R}^{768}$ as an aggregated representation[4] to fine-tune all the parameters end-to-end by adding a downstream auto-grading regression task. Specifically, for the fine-tuning with regression, we add a Feed-forward Neural network (FNN) with weight matrix $W \in \mathbb{R}^{1\times 768}$ and bias b, as well as a *Sigmoid* activation function σ to predict an essay score s, as shown in Formula 1.

$$\begin{aligned} FNN(h_{[CLS]}) &= Wh_{[CLS]} + b \\ s &= \sigma(FNN(h_{[CLS]})) \end{aligned} \tag{1}$$

We compute a standard regression loss Mean square error (MSE) with $h_{[CLS]}$ and W, as shown in Formula 2

$$MSE(s,t) = \frac{1}{l}\sum_{i=1}^{l}(s_i - t_i)^2 \tag{2}$$

where t are the gold scores of essays, and l is the number of samples.

[4] $BERT_{base}$ model is used in this study, whose transformer layers are 12 and hidden size is 768.

3.2 Using Self-attention for Retrieval of Key Topical Sentences

BERT consists of multiple layers of Transformers [20]. Self-attention is the most essential mechanism in transformers, which is used to calculate token-to-token weights in an input sequence. Specifically, input embeddings $[h_1, h_2, \ldots, h_n]$ corresponding to tokens from S are fed to the attention head. Matrix Q, K, V, referring to query, key and value are constructed. Attention head computes attention weights between all pairs of tokens through softmax of the dot product between the matrix Q and K, while the output of the attention head is the weighted sum of V, shown in Formula 3.

$$Attention(Q, K, V) = softmax(\frac{Q^T K}{\sqrt{d_k}})V \tag{3}$$

The attention weight between any two tokens t_i, t_j from the input sequence at layer l is:

$$\alpha_l(t_i, t_j) = softmax(\frac{q_i^T k_j}{\sqrt{d_k}}) \tag{4}$$

which can be interpreted as the degree of "importance" of the other token when calculating the new representation of the current token [15]. Inspired by this, we hypothesize that if the average attention scores between the tokens in a sentence and the keywords are higher than others, the sentence would be viewed as a more important sentence that affects the representations of the keywords, which could be regarded as the KTS. Thus, we calculate the attention scores between the sentence s and keywords at layer l according to the following formula:

$$Atten\_Sent(s, l) = \sum_{j=1}^{m} \frac{\sum_{i=1}^{n} \alpha_l(t_{s_i}, t_{k_j})}{n} \tag{5}$$

where n and m are the number of tokens in topical keywords and the sentence. We use NLTK[5] sentence tokenizer to split each student essay into sentences, and then rank the sentences according to the sentence attention scores in descending order. Considering BERT consists of 12 transformer layers, we investigate which layer/layers perform the best by calculating sentence attention scores from each layer ($Atten\_Sent(s, l)$), and all layers ($\sum_{l=1}^{12} Atten\_Sent(s, l)$) against human annotated KTS from student essays.

4 Experiments and Results

In this section, we first illustrate the experiment settings of training topic-aware BERT, related evaluation metrics, baseline models, and results. Further, we introduce how we annotate the KTS from student essays and evaluate our KTS-retrieving method.

[5] https://www.nltk.org/.

4.1 Topic-Aware BERT Training, Evaluation, and Results

Experiment Settings. We use only 5,152 essays from prompts 1, 2, and 7 in our study, while the previous state-of-art works deploy 12,978 essays from all prompts. To avoid the bias introduced by the dataset size difference, we replicate the experiments of selected previous state-of-the-art models with the same essays used in our study as baselines. Specifically, CNN-LSTM [8], LSTM-CNN-Att [9], and BERT [10] without topical information are used as baselines. We deploy 5-folds cross-validation to evaluate all models. In each fold, essay examples are split to train (60%), test (20%), validation (20%) dataset, which is in line with the standard experiment settings in the previous works [8,9]. We fine-tune BERT uncased base model[6] with keywords and essays as described in Sect. 3.1 for ten epochs, with a learning rate at 1e−5 and batch size of 10. All experiments are performed on an NVIDIA 1080Ti GPU. Same as previous works, we normalize the essay scores to be in the range of 0 and 1. We choose the best models based on the validation dataset and evaluate against the official metric Quadratic Weighted Kappa (QWK) scores for each essay prompt.

Evaluation Metric. Quadratic Weighted Kappa (QWK) is the official evaluation metric for the ASAP dataset. To calculate QWK, we first construct a matrix W, where i and j are essay grades assigned by humans and machines, and N is the number of possible grades. Another two matrices O (each element stands for the number of essays that receive grade i and j) and E (outer product of histogram vectors of i and j) are calculated as well. QWK is calculated as

$$W_{(i,j)} = \frac{(i-j)^2}{(N-1)^2} \qquad \kappa = 1 - \frac{\sum_{i,j} W_{i,j} O_{i,j}}{\sum_{i,j} W_{i,j} E_{i,j}} \tag{6}$$

Experiment Result. The QWK scores achieved by different models are illustrated in Table 3. Generally, due to having fewer training examples, the performance of the baseline models is negatively affected. For instance, CNN-LSTM-Att based model achieves QWK of 0.801 when training with essays from all prompts, falling to 0.788 while experimenting with essays from prompts 1, 2, 7. We also realized that some scenarios, such as the CNN-LSTM-att model with prompt 1 essays and BERT model with prompt 7 essays, benefit from fewer training examples. As topics and grading rubrics differ from prompt to prompt, we suspect that in some cases training with fewer prompts of less topic and grading rubric inconsistency, it could be easier to automatically learn AES features for non-topic-aware models than training with all prompts. We will conduct future work to verify this. Regarding topic-aware BERT, the performance is robust and competitive compared to all baseline models either with all prompts essays or experimented essays. Specifically, topic-aware BERT achieves the highest QWK of 0.702 on prompt 2 essays and is the second-best model when predicting prompt 1 and 7 essays scores with QWK of 0.822 and 0.818 respectively.

[6] https://huggingface.co/bert-base-uncased.

The topic-aware BERT outperforms all selected previous state-of-the-art models and achieves 0.781 when measuring average QWK as the evaluation metric.

Table 3. QWK scores of different models with essays from prompts 1, 2, 7 in our experiment. The numbers in the parenthesis indicate the QWK scores reported in the papers of selected models experimenting with all prompt essays. For example, concerning essays from prompt 1, CNN-LSTM achieves 0.789 in our experiments, while the reported QWK in the original paper is 0.821.

Models	Essay prompts			Avg QWK
	1	2	7	
CNN-LSTM	0.789 (0.821)	0.687 (0.688)	0.805 (0.808)	0.760 (0.772)
CNN-LSTM-Att	**0.825** (<u>0.822</u>)	0.658 (0.682)	0.788 (0.801)	0.757 (0.768)
BERT	0.814 (0.821)	<u>0.689</u> (0.678)	**0.820** (0.802)	0.774 (0.767)
Topic-ware BERT	<u>0.822</u>	**0.702**	<u>0.818</u>	**0.781**

4.2 Key Topical Sentence Retrieving Method Evaluation

Evaluation Metric and Result. As described in Sect. 3.2, we retrieve KTS through ranking sentences by *Atten_Sent* through each transformer layer. The ranking system of each layer could be regarded as an information retrieval system. Thus *Mean Average Precision* (MAP), which have been widely used to evaluate information retrieval systems [21], is also used in this work to evaluate the performance of KTS retrieved from different transformer layer/layers in topic-aware BERT. MAP is calculated as:

$$MAP = \frac{1}{Q}\sum_{j=1}^{Q}\frac{1}{m_j}\sum_{k=1}^{n} P@k \times rel@k$$
$$P@k = \frac{\{\text{key topical sentences}\} \cap \{\text{retrieved sentences}\}}{\{\text{retrieved sentences}\}} \tag{7}$$

where Q is the size of the evaluation essays (=60), m_j is the number of gold KTS. $P@k$ is the precision@k, and $rel@k$ refers to a relevance function that equals 1 if the sentence at rank k is a key topical sentence and equals 0 otherwise. We take the best model checkpoints for each essay prompt and calculate the MAP scores of different layer/layers. For comparison, we use a non-neural approach, *term frequency-inverse document frequency* (TF-IDF) [22], which extracts sentences based on the word frequencies as the baseline. The result, shown in Table 4, indicates that our KTS-retrieving method performs more robustly when dealing with argumentative essays. Specifically, for prompts 1 and 2, which are argumentative essays, layer 10 of topic-aware BERT achieves the highest MAP scores of 0.68 and 0.69. For prompt 7, which are narrative essays, the topic-aware BERT performs less competitively. We could recommend layer 10 of topic-aware BERT to extract key topical sentences for argumentative essays.

Table 4. MAP scores of different layer/layers on each essay prompt. Eg. Layer 10 achieved MAP of 0.68, 0.69 and 0.51 respectively for essays from prompt 1, 2, 7.

Prompt ID	Transformer layer/layers ID in Topic-aware BERT													Baseline
	1	2	3	4	5	6	7	8	9	10	11	12	All	TF-IDF
1	0.58	0.56	0.52	0.52	0.55	0.55	0.58	0.59	0.55	**0.68**	0.52	0.54	0.55	0.56
2	0.55	0.62	0.54	0.55	0.44	0.6	0.6	0.59	0.59	**0.69**	0.56	0.6	0.6	0.53
7	0.51	0.51	0.54	0.53	0.4	0.51	0.52	0.52	0.54	0.51	0.54	0.53	0.53	0.55

5 Conclusion, Limitations and Future Work

We have proposed the topic-aware BERT to link automated essay scoring and automated writing evaluation in this work. The experiments illustrate that by feeding both topical information and student essays, topic-aware BERT achieves solid and robust AES performance compared with various previous state-of-the-art methods. Moreover, by probing self-attention scores between topical keywords and sentences in student essays, the 10th layer of topic-aware BERT shows effectiveness at spotting students' key topical sentences, especially in argumentative essays. With a reliably predicted essay score, extracted key topical sentences will accelerate teachers' grading process and improve the transparency of the AES system, making the AES system less like a black box. We also realize some unresolved questions, such as the generalization of our approach on other datasets and keywords. Thus, we plan to address these limitations by evaluating our method with various datasets and keywords, as well as conducting an experimental study to see whether this approach could benefit teachers when conducting essay grading in practice.

References

1. Ke, Z., Ng, V.: Automated essay scoring: a survey of the state of the art. In: Proceedings of the International Joint Conference on Artificial Intelligence, pp. 6300–6308 (2019)
2. Hussein, M.A., Hassan, H.A., Nassef, M.: Automated language essay scoring systems: a literature review. PeerJ Comput. Sci. **5**, e208 (2019)
3. Smolentzov, A.: Automated essay scoring: scoring essays in Swedish. Dissertation (2013)
4. Eckes, T.: Introduction to Many-Facet Rasch Measurement: Analyzing and Evaluating Rater-Mediated Assessments. Peter Lang Publication Inc., New York (2015)
5. Kumar, V.S., Boulanger, D.: Automated essay scoring and the deep learning black box: how are rubric scores determined? Int. J. Artif. Intell. Educ. **31**, 538–584 (2021). https://doi.org/10.1007/s40593-020-00211-5
6. Attali, Y., Burstein, J.: Automated essay scoring with e-® v. 2.0. ETS Res. Rep. Ser. **2004**(2), i–21 (2004)
7. Rahimi, Z., Litman, D.J., Correnti, R., Matsumura, L.C., Wang, E., Kisa, Z.: Automatic scoring of an analytical response-to-text assessment. In: Trausan-Matu, S., Boyer, K.E., Crosby, M., Panourgia, K. (eds.) ITS 2014. LNCS, vol. 8474, pp. 601–610. Springer, Cham (2014). https://doi.org/10.1007/978-3-319-07221-0_76

8. Taghipour, K., Ng, H.T.: A neural approach to automated essay scoring. In: Proceedings of the Conference on Empirical Methods in Natural Language Processing, pp. 1882–1891 (2016)
9. Dong, F., Zhang, Y., Yang, J.: Attention-based recurrent convolutional neural network for automatic essay scoring. In: Proceedings of the 21st Conference on Computational Natural Language Learning (CoNLL 2017), pp. 153–162 (2017)
10. Devlin, J., Chang, M.W., Lee, K., Toutanova, K.: BERT: pre-training of deep bidirectional transformers for language understanding. In: Proceedings of the Annual Conference of the North American Chapter of the Association for Computational Linguistics: Human Language Technologies, pp. 4171–4186 (2019)
11. Yang, R., Cao, J., Wen, Z., Wu, Y., He, X.: Enhancing automated essay scoring performance via fine-tuning pre-trained language models with combination of regression and ranking. In: Findings of the Association for Computational Linguistics, pp. 1560–1569 (2020)
12. Woods, B., Adamson, D., Miel, S., Mayfield, E.: Formative essay feedback using predictive scoring models. In: Proceedings of the 23rd ACM SIGKDD International Conference on Knowledge Discovery and Data Mining, pp. 2071–2080 (2017)
13. Madnani, N., et al.: Writing mentor: self-regulated writing feedback for struggling writers. In: Proceedings of the 27th International Conference on Computational Linguistics: System Demonstrations, pp. 113–117 (2018)
14. Zhang, H., Litman, D.: Automated topical component extraction using neural network attention scores from source-based essay scoring. In: Proceedings of the 58th Annual Meeting of the Association for Computational Linguistics, pp. 8569–8584 (2020)
15. Clark, K., Khandelwal, U., Levy, O., Manning, C.: What does BERT look at? An analysis of BERT's attention, pp. 276–286 (2019)
16. Graff, G.: Clueless in Academe: How Schooling Obscures the Life of the Mind. Yale University Press, New Haven, CT (2003)
17. Hillocks, G., Jr.: Teaching Argument Writing: Supporting Claims with Relevant Evidence and Clear Reasoning. Heinemann, Portsmouth, NH (2011)
18. Kuhn, D.: Education for Thinking. Harvard University Press, Cambridge, MA (2005)
19. Newell, G., Beach, R., Smith, J., VanDerHeide, J.: Teaching and learning argumentative reading and writing: a review of research. Read. Res. Q. **46**, 273–304 (2011)
20. Vaswani, A., et al.: Attention is all you need. Adv. Neural Inf. Process. Syst. **30** (2017)
21. Manning, C.D., et al.: Evaluation in information retrieval. Introduction to Information Retrieval, pp. 151–175. Cambridge University Press, Cambridge (2008)
22. Wu, H.C., et al.: Interpreting TF-IDF term weights as making relevance decisions. ACM Trans. Inf. Syst. **26**(3), 13:1–13:37 (2008)

Automatic Educational Concept Extraction Using NLP

Xiu Li(✉), Jalal Nouri, Aron Henriksson, Martin Duneld, and Yongchao Wu

Stockholm University, NOD-Huset, Borgarfjordsgatan 12, 16455 Stockholm, Sweden
{xiu.li,jalal,aronhen,xmartin,yongchao.wu}@dsv.su.se

Abstract. Educational concepts are the core of teaching and learning. From the perspective of educational technology, concepts are essential meta-data, representative terms that can connect different learning materials, and are the foundation for many downstream tasks. Some studies on automatic concept extraction have been conducted, but there are no studies looking at the K-12 level and focused on the Swedish language. In this paper, we use a state-of-the-art Swedish BERT model to build an automatic concept extractor for the Biology subject using fine-annotated digital textbook data that cover all content for K-12. The model gives a recall measure of 72% and has the potential to be used in real-world settings for use cases that require high recall. Meanwhile, we investigate how input data features influence model performance and provide guidance on how to effectively use text data to achieve the optimal results when building a named entity recognition (NER) model.

Keywords: Concept extraction · NLP · BERT · Sequence model · NER

1 Introduction

Concepts are crucial pedagogical elements in education. As digitalization progresses in education, massive online learning materials become increasingly available. In turn, how to effectively use the unstructured learning materials in texts and connect them in order to give the best learning experiences has become a challenge. One way to address this challenge can be labeling concepts in learning materials as representative meta-data. In this way, we break down learning modules into micro-learning components in concepts and the learning process can be easily navigated through concept maps. Concept map as a tool of constructivist learning can enhance cognitive development and improve learning performance [1]. The learning status of an individual can be measured on the level of concept in the knowledge tracing model, which further enables adaptive learning [2, 3]. Learning materials are connected through concepts so that a content-based adaptive recommender is possible [4]. Therefore, extracting and annotating concepts for learning materials is the most fundamental task and essential from both a pedagogic perspective for student modeling and knowledge modeling [5] and a technical perspective to facilitate various downstream functionalities in an intelligent learning platform to serve in the teaching and learning processes.

M. Temperini et al. (Eds.): MIS4TEL 2022, LNNS 580, pp. 133–138, 2023.
https://doi.org/10.1007/978-3-031-20617-7_17

In the real world, most of the concept labeling work is done manually by the teacher or book author, which is highly time-consuming and therefore not optimal in the big data era. To automate the process, the approaches of building concept extraction models can be generally categorized into rule-based learning, supervised learning, and unsupervised learning. Rule-based techniques can be token/nominal chunk-concept alignment or dictionary look-up. A typical unsupervised learning approach is topic modeling, which uses topic word distribution and topic mining to extract concepts. Most neural networks and deep learning-based concept extraction systems are based on supervised learning. However, annotating large datasets requires huge labor and domain knowledge. To avoid tedious annotation, some experiments use back-of-the-book indexes as "proxy" to train the concept extraction model, adapted to the conventional paper books. Other large-scale projects such as DBpedia Spotlight use Wikipedia hyperlinks to build a system that can find mentions of DBpedia ontologies. The limitation with DBpedia Spotlight is that our use cases are specialized in educational concepts and Wikipedia hyperlinks can be of any kind.

In this paper, we introduce an NLP-based approach to automatically extract and annotate concepts from Swedish learning materials. Specifically, we implement a NER model using KB-BERT, a pre-trained language model for the Swedish language, and fine-tune it on 8 annotated biological digital textbooks. We choose Biology as a departure point for building the concept extractor, as it is a concept-rich subject. This is the first concept extractor using fine-annotated digital textbook data that cover all content for the subject Biology for K-12 using Swedish BERT model. During fine-tuning, we experiment with five variations of data input generated from the same texts and concept set, inspired by the results from recent research [6] on interpretable evaluation for the NER task with regards to how annotation features affect NER model performance. We achieved a promising biological concept extractor with 72% in recall and give our insights on how to effectively use text data to build an optimal NER model.

2 Data

The digital textbook data are in HTML format and the concepts are explicitly annotated through tags. There are two types of tags in the digital textbooks: i) a_tag: concepts with a corresponding word explanation in the database; ii) i_tag: concepts without word explanation and are shown in italic format. The concepts can be tagged multiple times if they are the core concepts to learn in certain sections according to the curriculum and teaching plan, but in other sections they are no longer tagged. Descriptive statistics of the original biological textbook data are shown in Table 1. We choose the widely used IOB (Inside, Outside, Beginning) tagging scheme for data pre-processing.

3 Methods

3.1 Model

NER is a sequence labeling task in information retrieval, which is suitable for concept extraction. BERT is a transformer-based model that uses deep bidirectional architectures and research has shown that it performs well compared to other models on

Table 1. Descriptive statistics of the original biological textbook data.

# Books	# Paragraphs	# Sentences	# Tokens	# A_tag	# I_tag
8	8272	26149	361112	a_tag: 6531 distinct: 3305	i_tag: 3295 distinct: 1980

NER tasks [7]. Besides, we prefer a BERT model pre-trained on corpora that include Wikipedia. Wikipedia as the largest online encyclopedia has a nature like textbook and the downstream task is educational concept extraction. Therefore, Wikipedia has more relevant data as training data than other popular corpora such as MC4, which are data from web crawling but do not include Wikipedia. Among the available Swedish BERT models, RoBERTa is trained on MC4 and Swedish ALBERT is in alpha version. BERT base is trained on the National Library of Sweden's (KB) collections that include books, news, government publications, Swedish Wikipedia, and internet forums [8]. Considering model stability and usability, we finally choose BERT base for our task.

3.2 Experiments

We implement five experiments to train KB-BERT NER models with five different methods for data input and annotation. The motivations are: i) seeking concept extraction models with good performance; ii) with the first experiment as baseline, we subsequently form new hypotheses based on research [6] on how data features (especially entity density, label consistency, and context) influence NER model performance; iii) investigate how data features for training, validation and test influence the results.

We use both a_tag and i_tag as gold concepts. Each paragraph is the preliminary input example and we split data into 70/20/10 for training/validation/testing. Table 2 shows data wrangling and input features for the five experiments. The percentage of annotated input represents how balanced the input data is. Entity density represents the percentage of annotated tokens. We calculate label consistency for each annotated concept by counting the number of times the concept is annotated as concept divided by the number of times the concept actually appears in the whole dataset. Figure 1 illustrates the distribution of label consistency for all the annotated concepts in the original dataset. Only 34% of the annotated concepts have 100% label consistency. The mean of label consistency for all annotated concepts is 55%.

Experiment 1. Traditional BERT model uses sentence-level input, i.e. one sentence in each training example. We therefore break paragraphs into sentences as input. 25% of the input sentences are annotated with at least one concept and 3% of tokens are annotated as concept words. We use this model as baseline.

Experiment 2. We experiment to examine the **hypothesis** that, with other settings the same, adding more context to each training example would improve the model performance. An intuitive idea here is with consideration to the fact that many complex concepts take the whole paragraph to illustrate. In order to fully use relevant sentences to the same concept to disambiguate entities, we decide to take paragraph-level data as input to the BERT model.

Table 2. Data wrangling and input features for the experiments. Sen. = Sentence, Par. = Paragraph.

	Input	Re-annotated?	% annotated inputs	Label consistency	Entity density
Exp. 1	Sen	No	25%	55%	3%
Exp. 2	Par	No	48%		
Exp. 3,4,5	Sen	Yes	82%	100%	15%

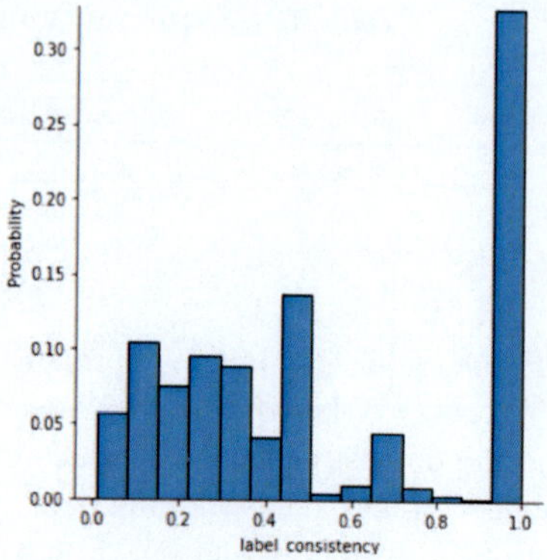

Fig. 1. Distribution of label consistency for annotated concepts in the original data.

Experiments 3, 4, 5. The concepts are only annotated in certain sections of the books as core concepts to learn but not annotated in other sections although they are mentioned. This results in low label consistency (55%) for the original dataset used in Experiments 1 and 2. In Experiments 3, 4, 5, we re-annotate all appearances of the concepts in the original data to examine the **hypothesis** that increasing label consistency improves model performance. We implement Experiments 3, 4, and 5 using re-annotated data (100% label consistency) for model training, but various data for model validation and testing. The data usage in these five experiments are described in Table 3. After the re-annotation, the percentage of annotated sentences increases from 25% to 82% and entity density increases from 3% to 15%.

4 Results

Table 4 shows the results of the five experiments. Experiment 1 in general gives a low performance, resulting in an F1 of 46%. It is in line with the conclusion in [6] that it is still challenging for contextualized pre-trained NER systems to handle entities with lower label consistency and lower entity density.

Using Experiment 1 as baseline, the **hypothesis** of Experiment 2 that adding more context to each training example would improve the model is **not approved** since the F score becomes slightly worse. This is in contrast with the results in [6]. The reason can be that [6] gradually tests how the number of added contextual sentences to the original sentence affects model performance, but how much connection between the original and added sentences is uncertain. However, sentences in a paragraph are naturally strongly correlated in content in our textbooks, which means our experiment has better settings here. Besides, the number of sentences in a paragraph is not fixed in Experiment 2 but [6] uses fixed numbers of contextual sentences. Furthermore, [6] uses different variations of LSTM models and we use BERT models.

Using re-annotated dataset means using data with high label consistency, higher entity density and more balanced input examples. When we compare the results of Experiment 1 with Experiments 3, 4, training the model on the re-annotated dataset gives a

Table 3. Data usage for model training, validation and testing in the experiments. *Experiment 2 uses paragraph instead of sentence as input.

	Training dataset	Validation dataset	Test dataset
Exp. 1	Original	Original	Original
Exp. 2*	Original	Original	Original
Exp. 3	Re-annotated	Original	Original
Exp. 4	Re-annotated	Re-annotated	Original
Exp. 5	Re-annotated	Re-annotated	Re-annotated

Table 4. Model evaluation of the five experiments.

	Precision	Recall	F1
Exp. 1	0.5978	0.3744	**0.4605**
Exp. 2	0.6053	0.3698	0.4591
Exp. 3	0.1530	0.7021	0.2512
Exp. 4	0.1546	**0.7175**	0.2544
Exp. 5	0.9033	0.8999	**0.9016**

significant boost to the recall to identify core concepts since the recall has increased dramatically from 37% to > 70%. It is expected that recall improves while precision degrades since the model trained on re-annotated data will likely predict more concepts while some of which are not labeled as such in the test set. Especially when we compare Experiments 1 and 5, where Experiment 5 uses re-annotated dataset for all model training, validation and testing while Experiment 1 uses only the original data, all evaluation metrics become significantly better with re-annotated data. The F1 score has increased from 46% to 90%. This indicates that label consistency has a big influence on model performance. The **hypothesis** that increasing label consistency improves model performance, especially in recall is **approved**. The model in Experiments 4 and 5 is promising to be deployed in production as a one-subject domain-specific concept extractor, especially for use cases that require high recall. Moreover, although Experiment 1 uses only the original dataset with lower label consistency and entity density, the F1 score outperforms Experiments 3, 4 which mix using original and re-annotated data. Similarly, Experiment 5 that uses only re-annotated data outperforms Experiments 3, 4. We hence conclude that the model gives the best performance in F1 when training, validating and testing the model with datasets that have similar features in label consistency and entity density.

5 Discussion

This paper described building a Bert-based Swedish educational concept extractor in the subject Biology for K-12. We designed five experiments in order to gain a better BERT model and at the same time investigated how data features correlate to model performance. We achieved a biological concept extractor model for the Swedish language with acceptable performance, especially for use cases that require high recall. The experiments and results provided guidance to researchers on how to work with text data to achieve optimal results with NER model. Our results approved the importance of label consistency for NER model. Label consistency and its distribution should always be checked first before modeling. If the data has low label consistency, we can consider re-annotating the dataset if our use cases are aiming at achieving high recall. In a concept extraction system, we care more about the performance of catching all the concepts the system should catch i.e. a good recall than precision. Extracting more educational concepts around the original core concepts for the same input text is pedagogically helpful for students to build connections between concepts. We cannot see clearly the effects of entity density in our experiments since entity density increased automatically with label consistency during re-annotation. Introducing richer contextual information did not improve concept extraction due to the different experiment settings and there are more decisive factors such as label consistency that dominate the model performance than larger contexts. However, to train, validate and use the model on data that have similar features plays a more important role than label consistency.

The concept extractor in this work is domain-specific and we aim to build a large-scale multi-subject educational concept extractor in the future. Besides, seeking the correlation between concepts in preparation to build the hierarchy of concept maps is also relevant and can be explored further.

References

1. Novak, J.D., Musonda, D.: A twelve-year longitudinal study of science concept learning. Am. Educ. Res. J. **28**, 117–153 (1991)
2. Thaker, K., Carvalho, P., Koedinger, K.: Comprehension factor analysis: modeling student's reading behaviour: accounting for reading practice in predicting students' learning in MOOCs. In: Proceedings of the 9th International Conference on Learning Analytics & Knowledge, pp. 111–115. ACM, Tempe AZ USA (2019)
3. Thaker, K., Huang, Y., Brusilovsky, P., Daqing, H.: Dynamic knowledge modeling with heterogeneous activities for adaptive textbooks. In: The 11th International Conference on Educational Data Mining, pp. 592–595. IEDMS, Buffalo, NY, USA (2018)
4. Tarus, J.K., Niu, Z., Mustafa, G.: Knowledge-based recommendation: a review of ontology-based recommender systems for e-learning. Artif. Intell. Rev. **50**(1), 21–48 (2017). https://doi.org/10.1007/s10462-017-9539-5
5. Chau, H., Labutov, I., Thaker, K., He, D., Brusilovsky, P.: Automatic concept extraction for domain and student modeling in adaptive textbooks. Int. J. Artif. Intell. Educ. **31**(4), 820–846 (2020). https://doi.org/10.1007/s40593-020-00207-1
6. Fu, J., Liu, P., Neubig, G.: Interpretable multi-dataset evaluation for named entity recognition. In: Proceedings of the 2020 Conference on Empirical Methods in Natural Language Processing (EMNLP), pp. 6058–6069. ACL (2020)
7. Devlin, J., Chang, M.-W., Lee, K., Toutanova, K.: BERT: pre-training of deep bidirectional transformers for language understanding. In: Proceedings of the 2019 Conference of the North American Chapter of the Association for Computational Linguistics: Human Language Technologies, vol. 1, pp. 4171–4186. ACL, Minneapolis, Minnesota (2019)
8. Malmsten, M., Börjeson, L., Haffenden, C.: Playing with Words at the National Library of Sweden -- Making a Swedish BERT. ArXiv200701658 Cs (2020)

Digital Environment for Literacy and Future Education. A Pilot Experience of Serious Game Co-design

Stefania Capogna[1], Giulia Cecchini[2], Maria Chiara De Angelis[2](✉), Vindice Deplano[3], Giovanni Di Gennaro[2], Michela Fiorese[3], and Angela Macrì[3]

[1] Link Campus University, Via del Casale di San Pio V, 44, 00165 Rome, Italy
s.capogna@unilink.it

[2] Roma TRE University, Via del Castro Pretorio, 20, 00185 Rome, Italy
{giulia.cecchini,mariachiara.deangelis, giovanni.digennaro}@uniroma3.it

[3] Entropy Knowledge Network, Via Asmara, 26, 00199 Rome, Italy
vindice.deplano@ext.entropykn.net, {michela.fiorese, angela.macri}@entropykn.net

Abstract. The quality of teaching and the digital skills of teachers and students are increasingly taking priority in contemporary society, especially if related to the need to fight against functional and emotional illiteracy and unequal access to education. These areas of intervention are the objective of the DIG4LIFE - *DIGital Environment for LIteracy and Future Education* research-action project, which is co-founded by the Erasmus+ programme.

The project team has created a Serious Game by translating the DigComp 2.1 framework into an interactive digital simulation to support teachers in the assessment and training of students' digital skills and digital maturity through a gamified learning strategy. DIG4LIFE Serious Game (SG) is the concrete result of a highly structured Co-Design process that involves teams of teachers in 6 different countries. SG, which has been created ad hoc for the project, offers high school teachers the opportunity to use an engaging tool with students that concretizes a methodological approach in line with nowadays educational needs and the intrinsic demand for digital maturity. Serious Games give the possibility of situating the content learned, thus allowing those involved in training/education/instruction to evaluate the level of knowledge, know-how and mindset with respect to the subject/topic dealt with, as well as to train digital skills and digital maturity. The paper describes the process of co-design of the DIG4LIFE Serious game, which effectively becomes an "object to think with" and concrete opportunities for co-design and collaboration between teachers, trainers and students.

Keywords: Serious game co-design · Teachers' 'professionalization · Digital competences

M. Temperini et al. (Eds.): MIS4TEL 2022, LNNS 580, pp. 139–148, 2023.
https://doi.org/10.1007/978-3-031-20617-7_18

1 Introduction

DIG4LIFE is an action research project, co-founded by the Erasmus+ programme, that aims to create the best conditions for sharing best practices in teaching digital skills by the innovation and dissemination of innovative tools, such as simulator and teaching editor for digital gamification environment [1].

DIG4LIFE has built a transnational partnership among universities, research centres, training institutions, schools, including various approaches and experiences in the development of digital Technology Learning Environment (TEL).

The main objective of DIG4LIFE is to contribute by research and innovation to the improvement of teaching quality and digital skills of educators and students from upper secondary and VET school, so as to fight against functional and emotional illiteracy and unequal access to education.

The project also supports the adoption of European frameworks on digital skills of educators, citizens and organizations, including the development and use of open educational resources, open textbooks and Open-Source educational software.

DIG4LIFE provides different actions strongly correlated with each other: 1) the definition of a theoretical framework for self-assessment to share the entire design of research and to clarify methodology, instruments, tools, expected results; 2) a self-assessment of teachers' digital skills based on the DigCompEdu and the PIAAC Online; 3) the release of an open digital self-assessment tool based on the DigCompEdu framework.

The third one regards the co-design of a Serious Game (SG) or the digital skills self-evaluation as systematic approaches and opportunities for the initial and continuous professional development of teachers, who will be able to develop effective, open and innovative digital education methods and pedagogies, as well as practical tools.

The release of an open serious game for the digital skills self-assessment consists of three main tasks: 1) trainers training and co-design of the blended learning path for teachers; 2) teachers training and co-design of the serious game episodes; 3) testing of the prototype by upper secondary school and VET students (Fig. 1).

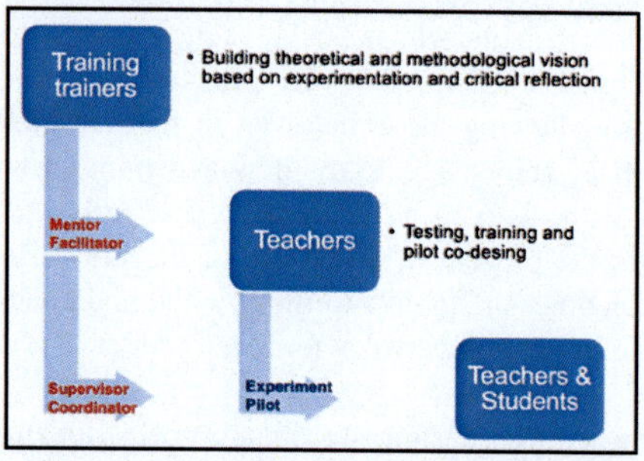

Fig. 1. DIG4LIFE Serious Game co-design process

The following paragraphs describes briefly the research question that inspired this work (Sect. 2), the co-design methodologies applied and the genesis of the serious game pilot (Sect. 3) that will be tested by students in the following months.

2 From Evaluation to Self-assessment

2.1 The Self-assessment as Connectors Between Teaching and Learning Process

For over twenty years, the evaluation has been at the center of debates and institutional investments to improve teaching quality, processes, and results. A new way of thinking about evaluation has developed under the weight of the critical issues that have always accompanied the evaluation theme in educational contexts. This new perspective focuses on *self-evaluation* to free the resources present in subjects and contexts, activating an actual development process and continuous improvement [2].

It has been proved that self-evaluation enhances motivation and progress in the learning process for teachers and students. Several studies have highlighted the positive relationship between teachers' self-assessment and professional growth [3–5].

Self-assessment tools connect teaching and learning process. Integrated with other personal growth strategies, they can improve teaching practices by a) increasing the teacher's awareness of teaching excellence levels, sense of efficacy and performance; b) helping the teacher in building improvement paths and in defining the necessary actions; c) facilitating communication between peers and the construction of professional communities of practice; d) stimulating constructive strategies to improve teaching effectiveness also through the influence of external change agents on teacher practice.

On the other hand, through the self-assessment strategy, students can evaluate their own work, reflect on their own learning and provide teachers with the perception of their learning. According to approaches based on the idea of *Self-directed learning* [6, 7], self-assessment methodology and tools can help both students and teachers to become aware of their strengths and weaknesses, set realistic goals for themselves and can define the stages and methods for achieving them, motivating their own learning process.

Based on this theoretical framework, the DIG4LIFE action research project wants to study the possible role that serious game could play as a *formative assessment tool*, to promote teachers' professional growth and student self-regulation and engagement, through a *self-reflection and co-design* working methodology that could impact positively on a) involvement and motivation; b) ubiquity and personalisation; c) creation of new ideas and knowledge; d) sharing and collaboration and e) increased experimentation.

2.2 A Co-design Methodology Through a TEL Experience

The origin of co-design dates back to the 1960s when trade unions in Scandinavia fought for *cooperative design*, the right of workers to co-design IT systems that impacted their work [8].

In the United States, starting from the 1970s, the term changed to *participatory design* and the need to involve end users in research is gaining more and more support. In the 80s Donald Norman published his famous book *Design of Everyday Things* in which he coined the term *user-centred design* and thus marked the transition to a design mentality [9].

The term has subsequently evolved into *human-centred design* [10] due to the attention given to a) focus on end users who brings their point of view into the design; b) multidisciplinary collaboration between people who bring their specific skills into the creative session; c) creation of a prototype that will be tested and improved by the beneficiaries.

In summary, among the advantages of co-design we point out the ability to respond in a targeted way to the needs of the reference target; to increase the interconnections between the participants in the co-design sessions; to reduce the time required for the development of ideas and promote innovation in a structured way.

In co-design, therefore, the activities are designed to facilitate sharing among the participants and align their ideas towards a common goal. At the same time this favours a continuous exchange and comparison between different points of view and requires negotiation skills in situations where diversity can become an obstacle to communication. During co-design sessions, participants feel involved in the process, thus developing a sense of deep responsibility towards the result [11–14].

Serious Game makes it possible to concretely implement a constructionist approach, which is one of the intrinsic methodological objectives of the project itself.

3 DIG4LIFE Serious Game: A Self-assessment Tool to Improve Digital Skills

3.1 DIG4LIFE Serious Game Prototype

DIG4LIFE Serious Game aims to translate a tool created for self-assessment (DigComp), into an interactive digital simulation that allows teachers to evaluate and train students' digital skills and digital maturity in a gamified way. The content of Serious game has been co-designed by the teachers of 6 different countries (Italy, Austria, Finland, Italy, Lithuania, Slovenia, Spain).

DIG4LIFE Serious Game has been inspired by three technological pillars that enhance creativity according to Papert [15, 16]:

- *Low Floors:* easy ways to get started for beginners
- *High ceilings:* works on increasingly sophisticated projects over time
- *Wide walls:* provide multiple paths from floor to ceiling. In this way, it will be possible to obtain highly personalised learning objects, improved learning through practice and greater motivation given the playful and emotional involvement of the participants.

These features are the best guarantee of success and transferability of the product to other teachers and students. Furthermore, the use of digital simulations allows teachers to put computational thinking and collaborative learning into practice. DIG4LIFE develops a methodology that, at the same time, defines the form and content around computational, creative thinking and digital maturity.

3.2 DIG4LIFE Serious Game: The Main Features

The DIG4LIFE challenge was to guide non-game designers in writing a good Serious Game story. National team were involved in co-writing a storyboard, which have three main functions:

1. to be a guide for designers to visualise the flow of the story and the crucial point (decisions to be made);
2. to have a detailed document that can be consulted by those who will realise the Serious Game and that provides instructions for its development (audio video contributions, editing and tests);
3. to be a useful repository for other countries involved in the DIG4LIFE project, also for peer review.

The storyboard of each episode took care of a general subject, two defined characters (protagonists) and two defined environments.

Concerning the subject, the SG six episodes took place in a digital future characterized by a very modern and technological smart e-society, in which characters must test a series of skills and competences to advance within the story, as well as to complete missions and challenges:

> *"Year 2050: a young boy aged 13 or over, is living in a campus with his peers. As we know, the school does not exist; there are no classrooms or lectures. The students live in the campus for about 3/5 years during which they receive assignments as concrete life experiences (informal training). Through experiences they acquire skills, knowledge and credits. The mentor interacts through a hologram, comments on the experience and assigns the score for passing the challenge, which means acquisition of competence. When the students reach a specific level of maturity (knowledge/skills) according to the evaluation of the mentor, they conclude the learning path."*

DIG4LIFE Serious Game has:

1. two protagonists and a robot in common with every episode in the series, as well as a specific character added by each country, if needed;
2. two environments in common (a, b), plus a specific one, if needed, for every episode (c):

   a Cafeteria, indoor tables, chairs, cups on table, tablets, computers, mobile phones, screens/pictures on the wall (to make objects clickable)
   b Campus room: computer, tablet, posters on the wall;
   c Hacker's "Cave": large tables with computers and pieces of computers. Laboratory neon lamps

Technical Specification

DIG4LIFE Serious games are made up of a proprietary architecture called "learning Brick, which provides a set of "prefabricated" modules in order to create a game with an assembly operation rather than with a technical development from scratch [17, 18].

Each game is built as a "learning object" that meets the Scorm interoperability standard: if inserted into a platform like Moodle, equipped with a compatible Learning Management System, it tracks all the main usage data for monitoring, evaluation, reporting and certification purposes: completion, times, scores, etc.). For maximum flexibility, these learning objects provide settings also to allow you to trace with xAPI standard [19]. With the xAPI standard, the eLearning content does not communicate directly with the LMS platform but with an intermediate level called Learning Record Store (LRS). This allows you to use the content even with a non-persistent connection and have a complete and updated tracking when the connection to the platform is re-established [20, 21]. Since only the classic languages of web applications are used (Html5, Javascript, Css), a PC or a mobile device equipped with a common browser and an internet connection are required. Alternatively, it is possible to develop versions for local use on personal computers (as portable Windows applications) or on Android smartphones and tablets (as installable apps).

3.3 Train the Teachers: The Training Path

DIG4LIFE Serious Game was created in collaboration with upper secondary and VET teachers as primary beneficiaries, who will use it as a tool for assessing their students' digital skills. Participatory design allows the creation of an innovative product tailored to the real needs of students.

The main co-design goals were to support and update teacher's digital skills, as defined in DigCompEdu [22], exploring and sharing pedagogical skills for educators.

The national teams were invited to join a learning experience based on a problem-based, gamified learning strategy, which created meaningful scenarios for them and consequently for students.

The lead partner assigned a digital competence to the national teachers' team. Each episode of the serious game includes the DigComp five expertise areas, declined into six digital competencies, as described in Fig. 2.

During the teacher training the following topics were addressed: competencies analysis based on the DigComp dimension, didactic design for digital simulations and game design.

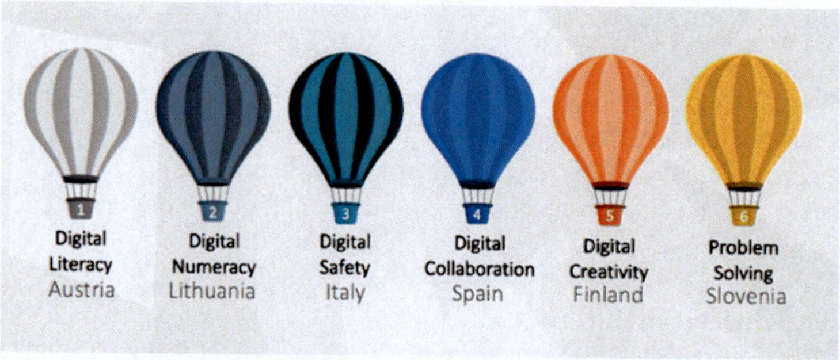

Fig. 2. DIG4LIFE Serious Game episodes assigned by national teams

The path was designed as a lab and was structured into two phases: a synchronous phase on Zoom platform and an asynchronous phase on the Moodle platform. The synchronous phase consisted of 10 workshops, 4 of which dedicated to the co-design of the storyboard, and 2 to the fine tuning of Serious Game.

The workshops aimed to:

- present project objectives and, in particular, the Serious Game tool;
- share the specific elements of Game design;
- introduce and support teachers in the storyboarding phase of Serious Game.

During the workshops teachers focused on the exploration of the objectives and the assigned DigComp competencies. The teachers with the help of the trainers analysed the sub-dimensions of the assigned competence, translating them into virtuous behaviours, from which to build subsequently the serious game storyboard.

They also started the activity in subgroups with the mentorship of the project team supporting and debriefing the storyboard (See Table 1).

Table 1. Workshop programme

Workshops	Subjects	Activities	Outcomes
1	*Serious game and co-design intro*	Play a serious game	A common experience of SG
2	*First step of co-design: hands on*	Working on skills and behaviour in subgroups	The SBS scheme
3	*Second step of co-design*	Drafting the plot of the story	Plot document
4	*Third step of co-design*	Creating the story and assigned scores/weights	Storyboard template
5	*Review serious game*		
6	*Fine tuning of the draft of Serious Game*		

The asynchronous phase provided a Moodle platform to give continuity between workshops. This phase focuses on the anticipation-follow up of content pillars, continuous feedback on what is produced and fine tuning of the outcomes. The management of the meetings was entrusted to a multidisciplinary team, consisting of project partners and trainers and/or tutors.

The first session begins with the presentation of the project's vision and strategy on how to transform the DigComp self-assessment questionnaire into an interactive digital simulation tool, allowing teachers to evaluate and develop students' digital maturity and skills. The teachers were accompanied to identify the major game design needs:

1. a subject, representing the idea, the narrative core of the story in which the fundamental components are described – environments, protagonists, context, and vicissitudes;
2. the script, or written elaboration of the subject that outlines the narrative structure of the story.

In this session, participants played the Italian SG pilot episode on *Digital Safety*[1] to better understand the main features are required. Practical experience helped to contextualize the theoretical explanation of the main features of SG.

The second training session got to the heart of the Serious Game design. The starting point was a detailed description of the chosen competence of the partner country, according to the provisions of the European Guidelines on DigComp [23].

Digital Safety is the DigComp competence assigned to the Italian group to build the pilot episode of DIG4LIFE Serious Game. In particular, the goal was to lead participants to a) define in detail the objective of the DIG4LIFE Serious Game episode on the assigned competence; b) describe the expected behaviours and specify the levels of mastery for the management of the assigned competence; c) identify and catalogue the three-four pillars representative of the assigned competence (pillars will be the internal variables of the Serious Game).

In the third training session, the activity was organized in two moments. The trainers presented the incipit of the story and requested the participants to modify or expand it; the teachers wrote the storytelling of the protagonists according to the main subject defined by the consortium, and the environments, characters, and the Skills-Behaviour Schema (SBS). The goal was to write the story starting from the main nodes and challenges that the protagonist had to overcome to "demonstrate" the level of management of the assigned competence.

Before starting the collaborative work in subgroups, the trainers explained the difference between plot and story. In order to write the story, the plot must be defined. The plot is linked to a specific and circumscribed event; while the story reveals how the characters react to this event (it is the emotional reaction to the choices made by the characters). In the plot teachers have to identify the following elements:

- the *exhibition:* information necessary to understand history (partially introduced by the Incipit);
- the *complications,* that trigger the "conflicts" to be resolved;
- the *climax:* the turning point in the history in which the "hero" has to resolve the situation;
- the *resolution:* the events that allow the closure of the story.

The participants, divided into subgroups, had to collaborate in the drafting of the story through a schema organized in three main steps: 1) incipit, constraints of the story, link to the SBS document (padlet); 2) first event (it was recommended to outline a story with 8/10 events; 3) the following event and so on.

The fourth session focused on completing all the elements of the story: steps, events, dialogues according to the SBS scheme proposed. The goal was to 1) finalise the work done before scoring; 2) make it consistent with skills and behaviours; 3) verify that the situations experienced by the protagonists really "measure" the skills assigned and the levels of mastery.

[1] The pilot episode on "Digital safety" is available at: https://www.entropylearningplatform.it/seriousgames/dig4life_it/pagine/lo.htm.

The last two workshops (5,6) were dedicated to review the Serious Game and debriefing of the episode.

The design of the training program has been planned in order to be reusable and to ensure that each iteration involves new members. The multiplication/transferability and the reusability of products are high because they are based on real needs and are produced on a cooperative basis with the target groups. Furthermore, the transfer of results is assured through the presence of partners who have knowledge and competencies for incorporating the learning/teaching methodology into the educational systems of the participating countries.

4 Conclusive Remarks

The DIG4LIFE project is now ongoing. The digital skills assessment tool DIG4LIFE Serious Game was co-designed and it's upload on the project Moodle platform.

The DIG4LIFE project is now ongoing. The digital skills assessment tool DIG4LIFE SG was co-designed and it's upload on the project Moodle platform. In Italy, the 25 teachers (distributed nationwide) who participated in the co-design of the SG, between May and September 2022 will complete the first testing phase of all the episodes, involving 321 high school students. Next step will be the experimental phase with the students of the schools involved in the partner countries.

In the DIG4LIFE serious game design is used a *flipped classroom methodology*. A *prescriptive and use-only approach* is abandoned to adopt a creative game design, an approach that puts co-design centre stage. In this framework the direct beneficiaries are "co-designers" and contribute with their real-world knowledge, perceptions and values across the entire game design process.

Co-creative game design creates a safe space for exploration and experimentation, supporting reflexive and metacognitive practices, opening opportunities to "think outside the box", admitting a multiplicity of representations and therefore a complex, multiform, and articulated image of reality. Co-creative serious game design allows to:

- *Create and not reproduce*: knowledge is 'created' by the mind rather than reproduced from external reality. Individuals are seen as 'builders of reality'.
- *Actively interact* with the environment, to experiment in comparison and construction the multiplicity and complexity of knowledge;
- *Collaborate and share* to negotiate and accept knowledge.

After testing, DIG4LIFE serious game will be available and free to be downloaded to give teachers and students the opportunity to self-assess their digital skills and to image appropriate strategies to became awareness citizens and empower their own digital maturity.

References

1. DIG4LIFE Project Homepage. http://dig4life.eu. Accessed 28 Apr 2022

2. Capogna, S.: La valutazione come strumento di empowerment organizzativo e professionale, pp. 98–117. RIV Rassegna Italiana di Valutazione, Franco Angeli (2020)
3. Festinger, L.: A theory of social comparison processes, in "Human Relations", 7. FLYNN FJ (2005), Identity orientation and forms of social exchange in the organizations. Acad. Manage. Rev. **30**(4), 737–750 (1954)
4. Peterson, K.D.: Teacher Evaluation: A Comprehensive Guide to New Directions and Practices. Corwin Press, Thousand Oaks, CA, London (2000)
5. Connelly, F.M., Clandinin, D.J.: Teachers as Curriculum Planners: Narratives of Experience. Teachers College Press, Ontario and OISE Press, New York (1988)
6. Kerka, S.: Self-directed learning: myths and realities. ERIC Clearinghouse on Adult, Career, and Vocational Education, Columbus, OH, USA (1994)
7. Mentz, E., Oosthuizen, I.: Self-Directed Learning Research: An Imperative for Transforming the Educational Landscape. AOSIS, Cape Town (2016)
8. Ehn, P., Kyng, M.: The collective resource approach to systems design. In: Bjerknes, G., Ehn, P., Kyng, M. (Eds.) Computers and Democracy - A Scandinavian Challenge, pp. 17–58. Gower Publishing (1987)
9. Norman, D.: The Design of Everyday Things. Basic Books, New York (1988)
10. Goodwin, K.: Designing for the Digital Age: How to Create Human-Centered Products and Services. Wiley, Indianapolis, IN (2009)
11. Krath, J., Schürmann, L., von Korflesch, H.F.O.: Revealing the theoretical basis of gamification: A systematic review and analysis of theory in research on gamification, serious games and game-based learning. Comput. Hum. Behav. **125**, 106963 (2021)
12. https://doi.org/10.1016/j.chb.2021.106963
13. Dichev, C., Dicheva, D.: Gamifying education: what is known, what is believed and what remains uncertain: a critical review. Int. J. Educ. Technol. High. Educ. **14**(1), 1–36 (2017). https://doi.org/10.1186/s41239-017-0042-5
14. Sailer, M., Homner, L.: The Gamification of Learning: a Meta-analysis. Educ. Psychol. Rev. **32**(1), 77–112 (2019). https://doi.org/10.1007/s10648-019-09498-w
15. Papert, S.A.: The Children's Machine: Rethinking School in the Age of the Computer. Basic Books, Pennsylvania State University (1993)
16. Papert, S.A.: Mindstorms: Children, Computers, and Powerful Ideas. Basic Books, Pennsylvania State University (1980)
17. Deplano, V.: Learning bricks: oggetti riusabili per simulazioni efficaci. Je-LKS **5**(2) (2009)
18. Deplano, V.: La simulazione come gioco e come modello di apprendimento. In: Castello, V., Pepe, D. (Eds.) Apprendimento e nuove tecnologie. Modelli e strumenti, Franco Angeli, Milano (2010)
19. Deplano, V.: I serious game nella didattica. Docete 21 (2020)
20. xAPI overview. https://xapi.com/overview/. Accessed 10 June 2022
21. xAPI solved and explained. https://xapi.com/?utm_source=google&utm_medium=natural_search. Accessed 10 June 2022
22. Redecker, C.: European framework for the digital competence of educators: DigCompEdu (No. JRC107466). Joint Research Centre (Seville site) (2017). https://doi.org/10.2760/178382
23. Carretero Gomez, S., Vuorikari, R., Punie, Y.: DigComp 2.1: the digital competence framework for citizens with eight proficiency levels and examples of use. EUR 28558 EN, Publications Office of the European Union, Luxembourg (2017). https://publications.jrc.ec.europa.eu/repository/handle/JRC106281. Accessed 28 Apr 2022

How Learnweb Can Support Science Education Research on Climate Change in Social Media

Apoorva Upadhyaya[1(✉)], Catharina Pfeiffer[2(✉)], Oleh Astappiev[1], Ivana Marenzi[1], Stefanie Lenzer[2], Andreas Nehring[2], and Marco Fisichella[1]

[1] L3S Research Center of Leibniz University Hannover, 30167 Hannover, Germany
{upadhyaya,astappiev,marenzi,mfisichella}@l3s.de
[2] Institute for Science Education, Leibniz University Hannover, 30167 Hannover, Germany
{pfeiffer,lenzer,nehring}@idn.uni-hannover.de

Abstract. We describe pilot study results evaluating how Learnweb's search and activity logs and eye tracking features can help to answer research questions of the SoMeCliCS project. We explored the extent to which students' information-seeking behavior can be tracked on a social media platform and how useful the results can be for science education.

Keywords: Social media · Climate change · Science education

1 Introduction

Social media (SM) have become a major information source and communication platform for young individuals and are increasingly important for formal education, also for science education. Many researchers have investigated to which extent scientific information, e.g. climate change (CC) information, is discussed in SM and which contents and features influence readers towards the topic [1]. Especially, scientific videos on SM platforms like YouTube are able to visualize complex phenomena and provide easy access to a variety of information. Therefore, they have benefits for science learning (e.g. [5]) and educators should foster students when they are seeking through SM environments in need of information. For decades, *literacies* were the educational approach to conceptualize the competencies, skills and behaviors students should develop in order to accomplish tasks like information evaluation (e.g. *information literacy* [6]), understanding media messages and sources (e.g. *media literacy* [6]) or solving scientific questions (*scientific literacy* [7]). Hence, a reconceptualization of some *literacies* according to rising challenges of online science learning is needed [2].

A main goal of the project "Social Media and Climate Change from the Perspective of Science Education" (SoMeCliCS) is to understand the influence of SM features on CC information-seeking behavior and to clarify the critical *literacies* in future studies. Due to the complexity of this goal, traditional investigation

M. Temperini et al. (Eds.): MIS4TEL 2022, LNNS 580, pp. 149–154, 2023.
https://doi.org/10.1007/978-3-031-20617-7_19

tools from science education research are not sufficient. It is necessary to supplement traditional methods with digital tools to investigate students' information-seeking behavior on SM like YouTube. To conclude on relevant SoMeCliCS *literacies* and determine educational implications for future studies, it is important to track students' perception of and interaction with information in SM.

Thus, it shall be investigated to which extent students' information-seeking behavior on SM platforms can, first of all, be tracked by Learnweb to further decide on the data usefulness for science education. Learnweb (Learnweb.l3s.uni-hannover.de/lw/) is a digital learning environment that provides users with a search interface for resource discovery and sharing across web services such as YouTube. The goal is to enable users to search for authentic resources from the web and reuse them in a learning context [3]. It provides features for organizing and sharing distributed resources with a group of people [4]. This paper illustrates a pilot study investigating how Learnweb was customized to support monitoring of information-seeking processes through pre-existing features and the new eye tracking functionality.

2 Theoretical Framework and Research Questions

In the future, SoMeCliCS researchers should be enabled to investigate (1) How students' search for CC information in the CC discourse on YouTube? (2) To which extent students perceive information and how do they interact with this information? and (3) How students organize the perceived information in the context of dismissive and consensus CC perspectives and video credibility and how do they critically reflect on their organization of video information? Therefore, the main research question of this pilot study is: 'To what extent do Learnweb features generate a suitable data output to further conclude on the SoMeCliCS research questions?'.

How students perceive multimedia information presented in videos (acustic, visual, textual) was discussed in many fundamental theories, also in the cognitive theory of multimedia learning (CTML) [8]. The CTML describes this according to three processes: selecting, organizing, and integrating [8]. In this study, YouTube as a video sharing platform is focused due to the above-mentioned video benefits for online science learning and students are also dealing with multimedia information. Therefore, research questions and survey details were mainly based on the CTML. Due to technical aspects of the study design and collectable data output, the three CTML processes were slightly modified and concretised as: (1) Selecting: searching for information (search terms, video URLs, search history), perceiving information (eye tracking data). (2) Organizing: interacting with information (likes/dislikes; YouTube comments; clicks; cursor movement; videos saved in private resources) (3) Integrating: critically reflecting on information (comments in Learnweb).

According to the main research question, authors refined Learnweb functionalities to allow more precise tracking of users' information-seeking behavior. In particular, the eye tracking shall complement the existing mouse tracking functionality. With eye tracking researchers are enabled to track video elements and

SM contents that are observed by the students (e.g. suggested videos, comments) and visualize what they were interacting with. In future SoMeCliCS studies, this additional data will supplement results from questionaires and interviews to reveal possible correlations between students' developed *literacies* and their actual information-seeking behavior.

3 Learnweb Components and Data Collection

In this section, we outline Learnweb components and the data gathered by each component while students performed the YouTube search.

Searching for Information: As the user searches for web resources related to CC, Learnweb tracks information such as the URLs accessed by the user, the timestamp of each action performed by the user, unique identifiers for each event, and the search queries used by the user to find the YouTube videos.

Perceiving Information: Eye tracking was implemented and is responsible for storing the data that is useful to perceive the information parameter. We integrated the webgazer.js webcam-based eye tracking library [9] into Learnweb, which uses the webcam to capture users' gaze in real-time with timestamps.

Interacting with Information: Learnweb captures user interaction with the help of a mouse activity tracker recording all mouse movements such as scrolling, clicking, typing, and cursor positions using Javascript. It also allows creation of specific folders where users can save the selected YouTube videos as their private resource *(accessible only to the creator of the folder).*

Organizing Information: Learnweb provides features to evaluate the credibility of the saved videos by providing a rating from '1' (not credible) to '5' (highly credible) and choosing a tag such as 'CC acceptance' or 'CC denial' in accordance with the users' perception of the video messages.

Critically Reflecting Information: A textbox is associated with each Learnweb resource where users can write comments to critically reflect their own credibility rating and their chosen tags.

4 Pilot Study Setting and Learnweb Tasks

Learnweb was tested with 17 participants (German university students). All students were informed about their rights and data privacy issues and asked to give informed consent to participate before data collection in agreement with the Data Protection Officer of our University. Then they completed a 25 min search query concerning general CC information in YouTube in accordance with the following instructions. After each YouTube search, feedback questions addressing benefits and limitations of the setting were asked. (1) The SoMeCliCS-project and Learnweb were briefly described to each student. Students then registered with pseudonym and password to the SoMeCliCS-course in Learnweb. (2) After the enrollment, students were led to the welcome page, where the search query described further steps. (3) A eye tracking re-calibration was performed for results below 65%. (4) The students started the search and saved YouTube videos that were recognized as 'relevant' for further evaluation of source credibility and video message by choosing a 'CC acceptance' tag or a 'CC denial' tag. (5) All videos were directly evaluated in the 'private resources' folder. (6) The students finally logged out and gave feedback about the study.

5 Results and Analysis

Searching for Information: in YouTube was tracked by saving URLs and search terms in each users' Learnweb search history. Students used general search terms (e.g. 'climate change reasons'), aspects of CC denial (e.g. 'climate change lie') or questions as search query (e.g. 'why climate change does not exist?'). Three students directly named the information source or a YouTube channel in their search terms (e.g., 'IPCC', 'oreskes', 'rezo'), showing an awareness of existing CC organizations, climate scientists and YouTubers. Furthermore, the search term 'oreskes climate change denial' combines a climate scientist with the topic of CC and the denial perspective, indicating that some students are already informed about CC discourses on SM while searching in YouTube.

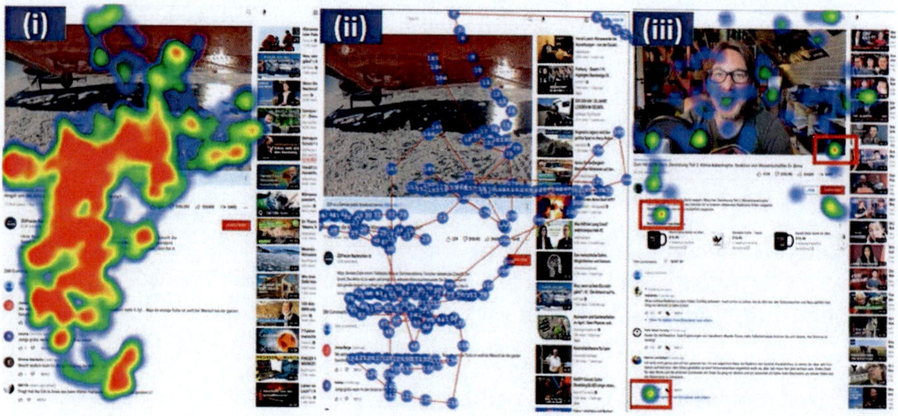

Fig. 1. (i) Eye tracking heatmap of student A's video. (ii) Eye tracking scanpath of student A's video. (iii) Mouse tracking heatmap of student's B video.

Perceiving Information: was monitored by generated heatmaps and scanpath sequences based on mouse and eye movements (Fig. 1). Heatmaps capture page elements that attract the most and least attention, while scanpaths can identify specific eye patterns. Figure 1(i) shows an eye tracking heatmap for student A of the visited YouTube video (youtube.com/watch?v=yliuiOVVtes), who viewed the content, checked the source, browsed the comments while briefly recognizing the suggested videos. The scanpath in Fig. 1(ii) helps further analyze student A's scanning behavior, such as reading processes (linear eye movement) or 'looking over' observations of the web page (searching for something), and shows a linear trend along video title, source description, and comments.

Interacting with Information: is determined as any kind of click/read/write actions related to SM features or the video. Learnweb captured any interaction through mouse events. Figure 1(iii) shows student B's mouse tracking heatmap of video (youtube.com/watch?v=OZA0ZS1hwdw) indicating that the student interacted with the web page by clicking on the 'SHOW MORE' button, 'View Full Screen' button in the movie player, focused on a comment near the 'Like' button and watched the suggested videos in the sidebar.

Organizing Information: required students to categorize the information perceived from the videos, relying on their prior knowledge, beliefs, and personal biases. Both credibility scores and tags could be based on personal factors, such as trust in science, experienced channels and YouTubers, and general worldview. Most students gave high credibility ratings to videos having trusted source or if they showed scientific experts using evidence. High credibility ratings were supported by comments such as 'message is consistent with my previous knowledge..' and 'trustworthy source (Terra X)'. Some students gave low ratings by commenting such as 'though video source was credible, but still video itself didn't have evidence..' and 'channel concerned with marketing its own books..'.

Critically Reflecting Information: demanded the students to write video comments in their 'private resources' folder for 'critically reflecting on the perceived information'. The comments provided information about students' video comprehension and their reasons for rating. Overall, comments referred to various SM features such as comments, source information, visible advertisements, or the appearance of the video itself that students relied on when personally organizing the video information. Some students commented on the comprehensibility of the video, e.g. 'explained simply' and 'Confusion caused by many numbers and calculations that are incomprehensible'.

6 Conclusions and Future Work

The pilot study examined whether Learnweb can support research on students' behavior when accessing CC content in YouTube. The data showed that Learnweb is able to track students' information-seeking behavior based on various parameters and provided deep insight into search terms and videos viewed, which were saved as private resources for other tasks. By analyzing the heatmaps, eye tracking provided an impression of the YouTube elements that students looked at while watching the videos. The scanpaths visualized students' perceptions of different content. Interactions with SM features in YouTube, such as likes/dislikes, comments, and suggested videos were tracked through clicks and cursor movements. The critical reflection of perceived video content were successfully tracked through the 'credibility rating' and 'tag' features, and students' comments in Learnweb made their categorizations understandable to science education researchers. Due to the small number of participants, the pilot study does not provide generalizable results. However, it was useful to figure out whether Learnweb, as a digital tool, can support science education research in the SoMeCliCS context. Moreover, the study revealed which additional methods are needed to complement these data. One methodological limitation is that eye tracking can generally visualize students' eye movements, which does not mean they actually perceived the content. Finding and analyzing trends in the scanpaths could help overcome this limitation. Another limitation is that we used webcam-based eye tracking library webgazer.js, which is fast, accessible and provided acceptable results in behavioral research [10], but webcam-based trackers are less precise and depend on lighting conditions.

Based on the students' feedback and the limitations of the webcam-based eye tracking, we plan to integrate webcam-based eye trackers and conduct a comparative analysis to select the best option for our research. We also plan to include a 'tutorial tour' in Learnweb and demo video as a guide for the tasks.

Acknowledgement. This work was partly funded by the SoMeCliCS project under the *Volkswagen Stiftung* and *Niedersächsisches Ministerium für Wissenschaft und Kultur*.

References

1. Lewandowsky, S., Cook, J., Fay, N., Gignac, G.E.: Science by social media: attitudes towards climate change are mediated by perceived social consensus. Mem. Cogn. **47**(8), 1445–1456 (2019). https://doi.org/10.3758/s13421-019-00948-y
2. Hoettecke, D., Allchin, D.: Reconceptualizing nature-of-science education in the age of social media. Sci. Educ. **104**(4), 641–666 (2020). https://doi.org/10.1002/sce.21575
3. Bortoluzzi, M., Marenzi, I.: WEB SEARCHES FOR LEARNING: how language teachers search for online resources. Lingue e Linguaggi **23** (2017)
4. Marenzi, I., Zerr, S., Abel, F., Nejdl, W.: Social sharing in Learnweb2. 0. Int. J. Continuing Eng. Educ. Life Long Learn. **19**(4), 276–290 (2009)
5. Rosenthal, S.: Media literacy, scientific literacy, and science videos in the internet. Front. Commun. **5**, 581585 (2020). https://doi.org/10.3389/fcomm.2020.581585
6. Koltay, T.: The media and the literacies: media literacy, information literacy, digital literacy. Media Cult. Soc. **33**(2), 211–221 (2011). https://doi.org/10.1177/0163443710393382
7. Bybee, R., McCrae, B., Laurie, R.: PISA 2006: an assessment of scientific literacy. J. Res. Sci. Teach. **46**(8), 865–883 (2009)
8. Mayer, R. E.: Cognitive theory of multimedia learning. In: The Cambridge Handbook of Multimedia Learning, Second Edition, pp. 43-71 (2014)
9. Papoutsaki, A., Sangkloy, P., Laskey, J., Daskalova, N., Huang, J., Hays, J.: Webgazer: scalable webcam eye tracking using user interactions. In: IJCAI (2016). https://par.nsf.gov/servlets/purl/10024076
10. Yang, X., Krajbich, I.: Webcam-based online eye-tracking for behavioral research. Judgm. Decis. Mak. **16**(6), 1486 (2021)

Open Government Data in Higher Education: A Multidisciplinary Innovation Teaching Experience

Iván Cantador[1](✉), J. Ignacio Criado[2], Laura Alcaide Muñoz[3], María E. Cortés-Cediel[4], and Irene Liarte[2]

[1] Escuela Politécnica Superior, Universidad Autónoma de Madrid, Campus de Cantoblanco, 28049 Madrid, Spain
ivan.cantador@uam.es

[2] Facultad de Derecho, Universidad Autónoma de Madrid, Campus de Cantoblanco, 28049 Madrid, Spain
{ignacio.criado,irene.liarte}@uam.es

[3] Facultad de Ciencias Económicas y Empresariales, Universidad de Granada, Campus de Cartuja, 18071 Granada, Spain
lauraam@ugr.es

[4] Facultad de Ciencias Políticas y Sociología, Universidad Complutense de Madrid, Campus de Somosaguas, 28223 Pozuelo de Alarcón, Spain
mcorte04@ucm.es

Abstract. In this paper, we present an innovative teaching project aimed to experiment with the use of open government data in higher education. Differently to previous studies on the topic, our project follows a multidisciplinary approach by conducting and evaluating several learning activities in degree subjects from distinct fields –political science, economics and finance, and computer engineering–, and targeting goals related to the development of professional and technological skills for understanding, accessing, analyzing and exploiting open data. Together with descriptions of these activities, which can be of inspiration for teachers having distinct backgrounds, we report results and valuable findings from a questionnaire-based study with students, providing insights about how to support effective teaching and learning using open data.

Keywords: Higher education · Digital competences · Interdisciplinary teaching · Learning innovation · Open data · Open government

1 Introduction

Nowadays, in the Big Data era, there is the need of data literacy, which entails learning how to read, process, analyze and argue with data [8]. It is related to the ability of searching and extracting knowledge from raw data [4] and, as stated by the OECD[1], it is a crucial skill that citizens have to acquire by 2030.

[1] http://www.oecd.org/education/2030-project/teaching-and-learning/learning.

M. Temperini et al. (Eds.): MIS4TEL 2022, LNNS 580, pp. 155–164, 2023.
https://doi.org/10.1007/978-3-031-20617-7_20

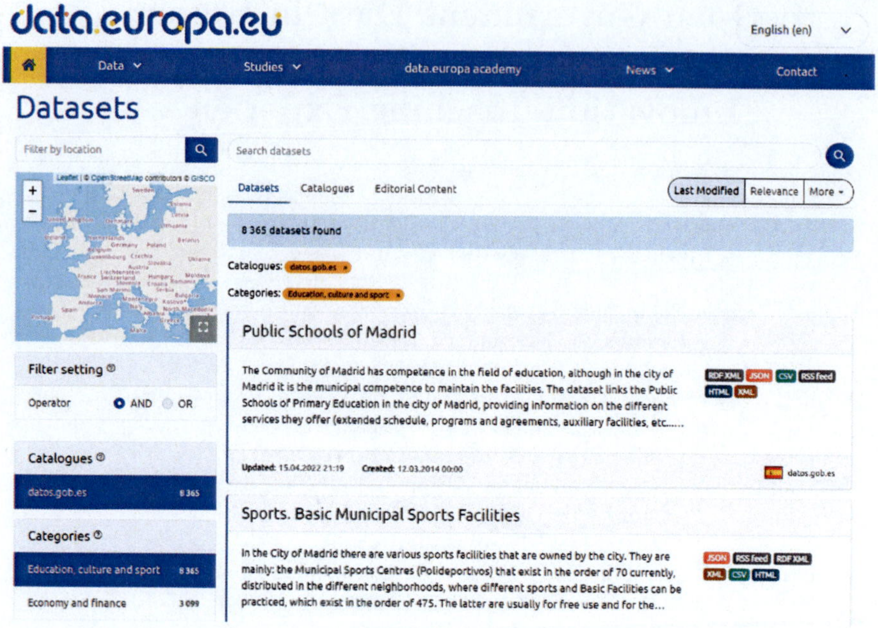

Fig. 1. Screenshot of the EU Open Data Portal, the official point of access to public data published by the EU institutions, agencies and other bodies. Its data collections are accessible via category menus, keyword-based search fields, and filtering criteria forms, among other mechanisms. Each collection can be provided in multiple electronic formats (e.g., CSV, XLS, XML, HTML, RDF, and JSON) and has associated metadata, such as a title, a description, and several tags.

In the educational context, students have to be guided in going beyond the passive inspection of results returned by a search engine, and in actively querying and looking for the data that best satisfy their information needs and allow informed decision-making [10]. Researchers have indeed identified a number of factors that should be taken into consideration in schools to develop suitable knowledge management, such as acquisition, learning, dissemination and transfer methods and technological solutions [2,14].

Among the existing Big Data sources, in this paper we focus on Open Government Data (OGD). As defined by the OECD,[2] OGD represents an initiative aimed to promote transparency, accountability and public value by making government data available to all. Facilitating the access and encouraging the use and free distribution of their datasets, governments foster the creation of business and the development of innovative, citizen-centric services [5].

OGD policies and portals of public administrations are thus one of the most recent trends of modernization and innovation in the public sector at international level. For instance, the EU Open Data portal (Fig. 1) provides access to

[2] https://www.oecd.org/gov/digital-government/open-government-data.htm.

more than 1.4 million datasets from 36 countries. In Spain, the open data portal of the General State Administration[3], controlled by the central government, contains over 57,000 datasets on a large number of domains, such as demography, employment, health, education, economy, culture, and natural environment, to name a few.

The openness, access, retrieval, processing and analysis of data are aspects of great interest and potential, particularly in the context of higher education [12]. Using open data in education entails a series of benefits. It allows designing practical and professional oriented courses. Serving as educational materials, it enables the development of student tasks aimed to address real-life problems [7]. This, as widely recognized in the Problem-based Learning literature, promotes the students' satisfaction and engagement during their learning process [6]. Additionally, open data have the potential to improve students' digital and data skills that are essential for future generations [15].

Despite these benefits, open data come up with a number of challenges [3]. For instance, there is an extended unfamiliarity with open data as a potential educational resource [12]. Instructors are not aware of the concept of open data and how they could integrate open data into their teaching. Even having knowledge on open data, they may have significant difficulties in finding relevant datasets [15]. Moreover, there is a generalized lack of skills; learners and educators may not have the literacies (digital and data skills) and resources (including time) to make open data useful for them [3].

The academic literature on education comprises reports of teaching experiences using open data in individual courses, e.g., databases [9], computer programming [7], and statistics [13], as well as interviews with instructors about the use of OGD in education [12]. These studies report isolated experiences, suggesting the need for further elaboration of comparative studies.

Differently to previous work, in this paper we present a multidisciplinary innovation teaching project aimed to implement and evaluate several OGD-driven learning activities for degree subjects in different fields, namely political science, economics and finance, and computer engineering. The activities do not only differ in the field of knowledge, but also in the way open data is used. Hence, we propose activities with different goals: understanding, accessing, analyzing and exploiting OGD. Moreover, the project includes an empirical study where students provided opinions and suggestions about the utility of OGD in learning, education and other contexts. For all the above, we believe our this work could be a valuable reference for educators interested in teaching with open data technologies, regardless of their background.

The remainder of the paper is organized as follows. Section 2 surveys related work on using OGD in higher education. Section 3 presents the multidisciplinary activities of our project. Next, Sect. 4 reports empirical results from the evaluation of such activities, and finally, Sect. 5 ends with some conclusions and future research lines.

[3] https://datos.gob.es.

2 Related Work

In the literature related to OGD and higher education, we can distinguish between two main lines of work: the *generation of educational resources as open data* [16] and the *use of open (government) data as educational resources in education* [1].

With respect to the use of OGD as educational resources, and focusing on **data exploitation goals**, Mazón et al. [9] performed a learning activity in a Databases course of a Computer Engineering degree where students had to propose an original scenario where different open data should be reused for a specific goal. Following a project-based learning methodology, the students had to design a relational database for managing the data in the envisioned scenario. The activity promoted a creative and entrepreneur attitude in students, as well as encouraged autonomous and lifelong learning. Besides, surveys made to students showed that reusing open data in a cooperative work context increased the students' motivation.

Also considering the exploitation of open data in the Computer Science field, Maksimenkova and Podbelskiy [7] conducted an activity in a computer programming course for 1st-year undergraduates in Software Engineering. The purpose of the activity was the implementation of computer programs able to read, process and visualize open data files. As stated by the authors, the utilization of open data was beneficial since it helped in making programming education "less artificial." However, the variety of format, complexity and quality of the open data collections was an issue that made the design and final success of the activity difficult.

Focusing on **data analysis goals**, Renuka et al. [11] designed a learning activity where Engineering students had to obtain and analyze open data to perform predefined studies related to the time evolution of urbanization factors, traffic intensity, and air quality levels in a city. The authors presented the results achieved by students, but did not report an evaluation of the activity. With similar learning goals, Rivera, Marazzi and Torres-Saavedra [13] conducted an activity for an introductory course in statistics where students had to analyze OGD collections to address real case studies, such as providing a metric of violence, analyzing demographics of hotel registrations, and studying variables related to types of properties in a country. As one of the main conclusions obtained from the activity, the authors claimed that the ubiquity of open data has the advantage of allowing teachers to easily 'locally adapt' lesson plans, integrating real data with context and purpose. They, by contrast, highlighted the lack of reliability in cases where it was not clear where the data came from.

From the literature survey, we can conclude that using open data has a number of benefits, such as increasing the students' engagement for addressing realistic problems, but entails additional efforts to instructors who have to carefully explore the data (format, complexity, quality, reliability) before using it in their classes. We observe that reported activities were done isolatedly in single subjects, and did not deal with the understanding of what open data are and how open data can be accessed. Our project, by contrast, follows a multidisciplinary approach considering several subjects and knowledge fields, multiple types of student tasks (e.g., individual oral presentations, collaborative and problem-based assignments, final degree works) and levels of OGD use: understanding, access, analysis and exploitation.

3 Multidisciplinary Activities

Our teaching innovation project aimed at promoting a culture of use of open data among instructors and students, developing learning materials related to open data, presenting the use of open government data portals as resources for analyzing the political-administrative, economic and legal realities, and developing professional and technological skills related to the access, analysis and exploitation of open data from public administrations.

In the context of higher education and through a multidisciplinary perspective –including political and economic sciences, as well as engineering studies–, we conducted several open data learning activities of different nature (i.e., writing reports, oral presentations, practical assignments, and final degree works) and methodologies (i.e., teacher-centered, student-centered, and project-based). To provide consistency, the activities were restricted to three OGD portals in Spain at national, regional and municipal level, from the governments of Spain[4], Andalusia[5] and Madrid[6], respectively.

We next briefly describe the activities, grouped by their usage focus of OGD, namely understanding, access, analysis, and exploitation.

3.1 Open Government Data Understanding

A first type of activities is focused on understanding what OGD are: their goals, benefits and challenges, formats and forms of publication, applications, case studies, etc. In this sense, students could for example be requested to search and survey references, and make written reports and oral presentations, explaining what they learned on certain issues and topics.

Regarding our project, in a 2nd-3rd-year subject on Organizational Theory and Public Administration of the Law and Political Science and Public Administration Double-Degree, at the Faculty of Law of Universidad Autónoma de Madrid (UAM), a group of students participated in an immersive experience oriented to promote their understanding of the political side of OGD portals, beyond the legal dimension. After some activities led by the instructors throughout a workshop, videos, and direct content observation, the students were encouraged to strengthen their understanding of OGD portals, both connecting vision (transparency) and voice (participation) dimensions of openness in the public sector.

---

4 https://datos.gob.es.

5 https://www.juntadeandalucia.es/datosabiertos.

6 https://datos.madrid.es.

Besides, for a Databases subject, a group of 3rd-year computer engineering students at the High Polytechnical School of UAM had to investigate advanced issues related to linked open data repositories (e.g., data formats and structures, storage systems, querying languages, linkage and publication schemas), focusing on the government domain. Students had to make individual seminars in the classroom, raising debates around the studied issues.

3.2 Open Government Data Access

A second type of learning activities is associated with the access to OGD, since both searching data collections and exploring the content of data files (in different formats) are tasks that require specific skills. In this context, students could be requested to find relevant data sources and collections, and extract from one or more data files particular answers or information for a given question or issue.

Within our project, in a 1st-year subject on Theory of Public Administration of the Management and Public Administration degree at the Faculty of Political Science and Sociology of Universidad Complutense de Madrid (UCM), a group of students participated in a number of workshops where they had to find, collect and present institutional information from open data portals in order to evaluate the levels of governance, transparency and accountability achieved by representative governments.

Moreover, two 4th-year computer engineering students at UAM did their final degree work on open government access problems. Each work entailed a brief survey of literature, design, implementation and evaluation of an application software, writing a report, and making an oral presentation for an academic committee. Specifically, the works consisted of a couple of intelligent dialog agents (chatbots) to assist on the access to OGD and e-participatory budgeting content through formal queries built via natural language conversations.

3.3 Open Government Data Analysis

A third type of learning activities comprises any kind of task involved in the analysis of OGD, namely the processing, integration and filtering of data, the use of statistical metrics and methods on selected data, and the application of visualization techniques and tools on data and analysis results. Students thus could be requested to perform a wide array of assignments aimed to analyze real-world phenomena.

In particular, our project encompassed a classroom assignment where the 3rd year accounting and finance students at the Faculty of Business Studies at Universidad de Granada (UGR) had to analyze main aspects of OGD platforms: data catalog, accessibility and visualization, and citizen participation. To do this, the teacher offered the students a statement explaining the above aspects, and allowed them to choose two of the three aspects. Students had to understand differences between analyzed platforms, explore real cases and that they could provide critical opinion on the state of this type of initiatives, compared to the disclosure of information in the private sector (more analyzed by them).

Also with data analysis purposes, a final degree work by a 4th-year computer engineering student at UAM aimed to develop a computer program to measure and analyze controversy on citizen debates (publicly available as open data) from an e-participation platform of Madrid City Council.

3.4 Open Government Data Exploitation

A fourth and final type of learning activities concerns the exploitation of (processed) OGD for a target purpose or application. Students could thus be requested to perform (complex) solutions to address real-world decision-making problems.

Within this category of activities, our project included a cooperative project-based activity in the Databases subject of the 3rd year of the Computer Engineering degree at UAM, where students had to design and build from scratch a relational database with OGD collected from the web. The application domain of the database built by each student team was freely chosen.

The project also comprised a final degree work at UAM, where a 4th-year computer engineering student implemented and evaluated a number of recommender systems that exploited OGD to suggest Madrid residents with citizen proposals that may be of their interest, according to both topic and location information. Hence, residents could be informed about city problems and proposed solutions. Moreover, relevant proposals may be supported by residents in the participatory budgeting processes of the city.

4 Empirical Results

Each of the classroom activities presented in Sect. 4 was conducted by a group of students. In all cases, after having done an activity, students were requested to voluntarily participate in our study by filling out an online questionnaire. We received feedback from 182 students (60% female, 40% male) with ages ranging from 18 to 26 years old (only 5% older than 23 years old). The questionnaire was composed of 5-point Likert scale questions aimed to evaluate the students' satisfaction with the activities and opinion about OGD. The questionnaire also had some questions for open comments and suggestions.

Figure 2 shows the distributions of scores given by the students on their perception about the utility of open data for doing academic work, complementing their study, enhancing their learning, and pursuing other purposes. Several interesting insights can be derived from them by considering the students' academic fields: computer engineering, economics and finance, and political science.

As one would expect, political science students are those that found OGD collections most useful for studying and doing academic work in their degree subjects. In this sense, some political science, and economics and finance students commented on the value of OGD, motivated by the possibility of performing data-driven sociocultural, political and economic analysis. Computer engineering students, by contrast, expressed a moderate opinion about the potential utility

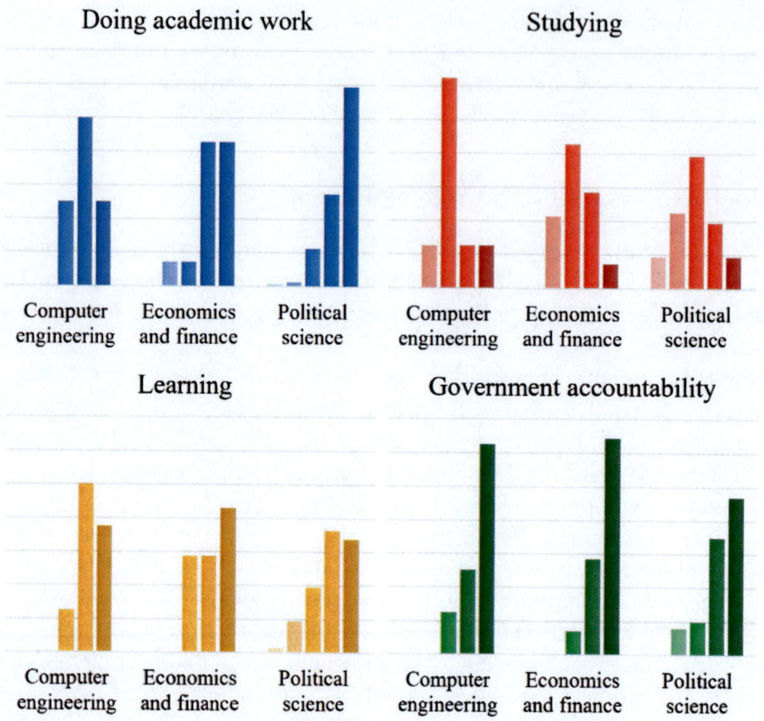

Fig. 2. Distributions of 5-point Likert scores (1 = very unuseful, 5 = very useful) given by the students about the perceived utility of using OGD for different purposes. The y-axis ranges from 0% to 70%, in intervals of 10%.

of open data for their studies. This may be due to the fact that their curriculum planning puts much more emphasis on learning algorithms and computer programming than data analysis and exploitation.

Nonetheless, interestingly, computer engineering students perceived open government as very useful for enhancing their learning. With this respect, some students mentioned the possibility of exploiting open data to build real-world databases and software applications, and to test machine learning methods on realistic prediction and classification tasks.

In general, regardless of their academic fields, students tended to give the highest score to the possibility of exploiting OGD for checking government accountability. Political science students were the most critical in this regard.

Finally, through open comments, students suggested the use of OGD for other (non-academic) purposes, such as verifying information published in news and social media, and ensuring government transparency. Here, one of the most interesting ideas that our activities raised was the opportunity of open data to tackle fake news and promote fact-checking of information, mostly in social media and algorithmic environments.

As reported through the questionnaire, 85% of the students were not aware of OGD portals before the activities. However, most of the students were satisfied or very satisfied with the activities (81%) and considered them as useful or very useful for their progress within the degree subjects (65%). This is another important conclusion of our study, as OGD portals comprise an asset of public sector organizations to engage with citizens, and promote trust and confidence in public institutions, including compulsory and higher education studies.

5 Conclusions

The academic literature recognizes the potential benefits of open (government) data as learning resources, but it also compiles a number of challenges that should be faced for their effective use; among them, the lack of teachers' awareness of the concept of open data, and how they can integrate open data in their subjects and classes.

Complementing published reports of individual experiences in particular higher education courses and surveys done with instructors, in this paper we have presented an innovative teaching project where distinct learning activities were done in subjects of different academic fields –political science, economics and finance, and computer engineering– at various levels of studying open data: understanding, access, analysis and exploitation.

Together with the categorized examples of activities that can be of inspiration for teachers having distinct backgrounds, we also have reported valuable findings obtained from a questionnaire-based multidisciplinary evaluation with students. The conducted experience showed the need for making changes on curriculum planning to develop or reinforce competencies on data processing and analysis, and for giving motivations to use open government data in daily life tasks, such as verifying the veracity of information published in news and social media, and making research works.

We plan to explore alternatives to the teacher-centered, student-centered and project-based learning methodologies followed in our project. Specifically, we want to perform new activities applying cooperative, flipped classroom, and challenge-based learning methodologies, as well as gamification mechanics (see examples given by Saddiqa et al. [15]). The activities could also be done in postgraduate studies, providing MSc and PhD students with opportunities to use open government data in their researches.

Our work may be of interest for scholars in different areas of knowledge and practitioners in the public sector. In the first case, colleagues in higher education institutions might find our cases and experiences inspirational for their own learning environments, including different areas of humanities, social and natural sciences, or engineering. In the second case, practitioners from public organizations may expand their approaches to OGD reutilization. In this context, public managers should include educational purposes in their portfolio of potential uses of their OGD portals, opening up a new area of interest among instructors and students at all levels.

Acknowledgements. This work was supported by Universidad Autónoma de Madrid (D_008.21_INN) and the Spanish Ministry of Science and Innovation (PID2019-108965GB-I00).

References

1. Atenas, J., Havemann, L., Priego, E.: Open data as open educational resources: towards transversal skills and global citizenship. Open Praxis **7**(4), 377–389 (2015)
2. Chu, K.W., Wang, M., Yuen, A.H.: Implementing knowledge management in school environment: teachers' perception. Knowl. Manag. E-Learn. **3**(2), 139–152 (2011)
3. Coughlan, T.: The use of open data as a material for learning. Edu. Tech. Res. Dev. **68**(1), 383–411 (2020)
4. De Donato, R., Garofalo, M., Malandrino, D., Pellegrino, M.A., Petta, A.: Education Meets Knowledge Graphs for the Knowledge Management. In: Kubincová, Z., Lancia, L., Popescu, E., Nakayama, M., Scarano, V., Gil, A.B. (eds.) MIS4TEL 2020. AISC, vol. 1236, pp. 272–280. Springer, Cham (2021). https://doi.org/10.1007/978-3-030-52287-2_28
5. Hardy, K., Maurushat, A.: Opening up government data for big data analysis and public benefit. Comput. Law Secur. Rev. **33**(1), 30–37 (2017)
6. Hung, W., Jonassen, D.H., Liu, R., et al.: Problem-based learning. In: Handbook of Research on Educational Communications and Technology **3**(1), 485–506 (2008)
7. Maksimenkova, O., Podbelskiy, V.: On practice of using open data in construction of training and assessment tasks for programming courses. In: 10th International Conference on Computer Science & Education, pp. 233–236. IEEE (2015)
8. Mandinach, E.B., Gummer, E.S.: A systemic view of implementing data literacy in educator preparation. Educ. Res. **42**(1), 30–37 (2013)
9. Mazón, J.N., Lloret, E., Gómez, E., Aguilar, A., Mingot, I., Pérez, E., Quereda, L.: Reusing open data for learning database design. In: 2014 International Symposium on Computers in Education, pp. 59–64. IEEE (2014)
10. Petrides, L.A., Nodine, T.R.: Knowledge management in education: defining the landscape (2003)
11. Renuka, T., Chitra, C., Pranesha, T.S., Dhanya, G., Shivkumar, M.: Open data usage by undergraduate students. In: 5th IEEE International Conference on MOOCs, Innovation and Technology in Education, pp. 46–51. IEEE (2017)
12. Rivas-Rebaque, B., Gértrudix-Barrio, F., de Cisneros de Britto, J.C.: La percepción del docente universitario ante el uso y valor de los datos abiertos. Educación XX1 **22**(2), 141–163 (2019)
13. Rivera, R., Marazzi, M., Torres-Saavedra, P.A.: Incorporating open data into introductory courses in statistics. J. Stat. Educ. **27**(3), 198–207 (2019)
14. Rodrigues, L.L., Pai, R.: Preparation and validation of KM measurement instrument: an empirical study in educational and IT sectors. In: Knowledge Management: Nurturing Culture, Innovation, and Technology, pp. 583–593. World Scientific (2005)
15. Saddiqa, M., Magnussen, R., Larsen, B., Pedersen, J.M.: Open Data Interface (ODI) for secondary school education. Comput. Educ. **174**, 104294 (2021)
16. Zablith, F., Fernandez, M., Rowe, M.: The OU linked open data: production and consumption. In: García-Castro, R., Fensel, D., Antoniou, G. (eds.) ESWC 2011. LNCS, vol. 7117, pp. 35–49. Springer, Heidelberg (2012). https://doi.org/10.1007/978-3-642-25953-1_4

Design and Computational Thinking with IoTgo: What Teachers Think

Andrea Bonani[1], Rosella Gennari[2], Alessandra Melonio[3](✉), and Mehdi Rizvi[4]

[1] Direzione Istruzione e Formazione italiana, via del Ronco 2, 39100 Bolzano, Italy
andrea.bonani@scuola.alto-adige.it
[2] Free University of Bozen-Bolzano, Piazza Domenicani 3, 39100 Bolzano, Italy
gennari@inf.unibz.it
[3] Ca' Foscari University of Venice, Via Torino 155, Mestre, Venice, Italy
alessandra.melonio@unive.it
[4] Politecnico di Milano, Via Ponzio, 34/5, 20133 Milan, Italy
syedmehdi.rizvi@polimi.it
https://www.researchgate.net/profile/Andrea-Bonani,
http://www.inf.unibz.it, https://www.unive.it/data/persone/25405996,
https://www.deib.polimi.it/eng/people/details/1636220

Abstract. Computational and design thinking are orthogonal and complementary ways of thinking, which are fundamental for nowadays' learners and yet taught in isolation. Teachers' understanding of them can be a barrier to their introduction. This paper reports on an intervention for primary- and secondary-school teachers, introducing them to both forms of thinking through hands-on laboratories, revolving around the IoTgo game-based toolkit. Teachers' ideas of computational and design thinking were investigated with a questionnaire before and after the intervention. Their answers suggest that the intervention was effective and indicate future work related to computational and design thinking.

Keywords: Design · Design thinking · Computational thinking · Smart thing · IoT · Teacher · Study

1 Introduction

Computational and design thinking are specific ways of thinking, rooted in a large body of knowledge and expertise in design and computer science, respectively, and useful in other fields. The ontological analysis of the two ways of thinking by Kelly and Gero suggests that "design thinking and computational thinking are processes that are ontological mirror images of each other, and are the two processes by which thinkers address problems" so that "thinkers can

This work is supported by the project SNaP of the Free University of Bozen-Bolzano and by DAIS - Ca'Foscari University of Venice within the IRIDE program.

M. Temperini et al. (Eds.): MIS4TEL 2022, LNNS 580, pp. 165–174, 2023.
https://doi.org/10.1007/978-3-031-20617-7_21

move fluently between the two". However, they are rarely employed together and even less taught together [11].

In fact, although both are considered fundamental 21st-century skills for learners, they are usually taught in isolation one from the other. Teachers' beliefs can be a barrier to their introduction in school settings, as Slotta et al. found in their investigation concerning the teaching of design and computational thinking with prospective teachers [19]. Relevant outcomes of their investigation for this paper are as follows: (1) teaching strategies which involve both forms of thinking employ project-based or design-centred approaches, (2) and, whereas a wide range of toolkits support computational thinking, few consider or encompass design thinking. They conclude that teacher support materials should explicit connect the two forms of thinking and engage learners through hands-on activities. This paper picks up their recommendation, and the one by Kelly and Gero: "given that these two forms of thinking are complementary ways of approaching problems, they might be taught in a way that emphasizes this relationship".

This paper presents an intervention for Italian in-service teachers from primary and secondary schools, concerning computational and design thinking. It engaged 32 of them in a one-month-long intervention. It introduced them to both forms of thinking through laboratories, with gamified IoTgo phygital toolkit that teachers can use at school on their own. This paper starts by presenting the most relevant background concerning interventions for teachers related to design thinking or computational thinking. Then it outlines the design of the intervention, sketching how the IoTgo toolkit was employed therein. Next the paper reports on the results of a study concerning participant teachers' ideas of the two forms of thinking, before and after the intervention. The results of the study are discussed in the conclusions to the paper.

2 Background

2.1 Design Thinking

Design thinking can be defined as a method of problem-solving that help people generate collaboratively novel solutions to open-ended, unstructured or ill-defined problems, which are to be understood as situated in a context [12]. However, the process is goal-oriented, constrained, and the resolution is specific and depends upon a designer's understanding of the situation [8]. In spite of its many definitions, a typical design thinking process includes the following stages: empathising with the problem/situation and people part of it, then defining and ideating a specific solution for a given goal, prototyping and testing it [9,13].

In recent years, design thinking has been adopted in educational settings, especially by the Maker movement [3]. In their view, educating children to think and act like designers helps them face difficult real-life situations and find "a solution" by themselves [15,18]. Much of past research on design thinking for education purposes focused on the potential benefits for learners [1]. Little research seems to investigate educators' ideas, perceptions or experience with design thinking [16]. In particular, the study by Hennessey et al. assessed teachers' impression of applying design thinking in their curricula [10]. Teachers reported positive impressions about its use for eliciting collaboration among students. However teachers also raised issues, especially concerning how to successfully make design thinking part of their curricula, e.g., specific toolkits or guidelines.

2.2 Computational Thinking

Computational thinking is considered as a set of skills and processes that enable students to face computational problems, e.g., problems which could be programmed and resolved with a computing machine [20]. Problems that require computational thinking are "typically recurrent problems, problems that either occur in many places or recur within the same place", and "the solutions provided by computational thinking aim to be generally applicable" [11].

That said, there is no single operational definition of computational thinking. However, the majority of them include the following stages: firstly abstraction of details, then pattern recognition for finding a general resolution, decomposition for breaking the problem or resolution into smaller more manageable parts, algorithms and programs to develop the resolution with computers.

Computational thinking has been widely advocated as a key component for education and teachers' training is considered a crucial factor for bringing it into school. Recently, researchers have investigated teachers' perceptions or understanding of computational thinking, their relation with smart technologies and their usage in class.

3 Study Design

3.1 Research Goal and Questionnaire

The research goal was to investigate teachers' ideas of computational thinking and design thinking, before and after the intervention, so as to assess possible effects of this on teachers' ideas.

An ad-hoc pre-post questionnaire was created, based on available ones in the literature, e.g., [2,19]. It was divided into a closed-format part and an open-format part. The closed-format part asked teachers to use a 5-point Likert scale and assess 3 groups of statements (Q1–Q3), each concerning either computational thinking (_CT) or design thinking (_DT). Statements to assess are reported in Table 1. Their order was randomised when the questionnaire was administered. Open-format questions completed the questionnaire, similar to those in the paper by Corradini et al., which investigated teachers' understanding of computational thinking [2]. Relevant open-format questions for this paper are as follows: (1) fill in freely "in my view, design thinking is...". (2) fill in freely "in my view, computational thinking is...".

The questionnaire was administered the first day of the intervention (Day 1), after presenting teachers definitions of computational and design thinking, with

Table 1. Closed-format part of the pre-post questionnaire for teachers, with three groups of statements (Q1–Q3) to be assessed on a 5-point Likert scale

Item	Statement to assess
Q1_CT	Computational thinking is useful in technical-scientific subjects
Q1_DT	Design thinking is useful in technical-scientific subjects
Q2_CT	Computational thinking is fundamental in today's society
Q2_DT	Design thinking is fundamental in today's society
Q3_CT	Computational thinking is useful in humanistic and artistic-musical subjects
Q3_DT	Design thinking is useful in humanistic and artistic-musical subjects

companion examples. The questionnaire was again administered at the end of the fourth day of the intervention (Day 4), after teachers had experienced the laboratories of the intervention.

3.2 Participants and Setting

Participants were 12 primary-school teachers and 20 secondary-school teachers. The intervention was held in the computer room of a secondary school, so that each teacher had a computer, a micro:bit physical-computing board and related devices (e.g., buttons, LEDs), besides internet access.

Notice that, in Italy, primary school teachers can teach all subjects. However, the 12 primary-school teachers participating in the intervention usually teach maths, science or technology related subjects, except one who is a support teacher for special-needs pupils. Out of all 20 secondary-school teachers of the study, 19 teach mathematics, science or technology, and one teaches art. Briefly, 96% of participants taught maths, science or technology subjects at school.

3.3 IoTgo Material

IoTgo is a phygital toolkit with game-boards and cards, besides digital tools, e.g., [5]. Its boards progressively and tangibly guide people through the creation of smart things and reflections around them, moving them fluidly across design and computational thinking, as recommended by Kelly and Gero [11]. In particular, the physical and cloud boards immerse people in a context via a mini-story, and help them choose things to make smart for certain personas and goals (presented as missions) as in design thinking (e.g., by empathising with personas). Next, they guide people to develop their ideas of smart things by means of physical inputs (e.g., touch sensors, buttons), physical outputs (e.g., LEDs, speakers), and by connecting them to cloud services via IoT communication, and reason as in computational thinking (e.g., by decomposing smart-thing ideas with given patterns, abstracting away details). Figure 1 shows a filled-in physical board.

The IoTgo toolkit includes ad-hoc hardware and software, namely, a scanner and a web app for: (1) reading cards and automatically generating programs

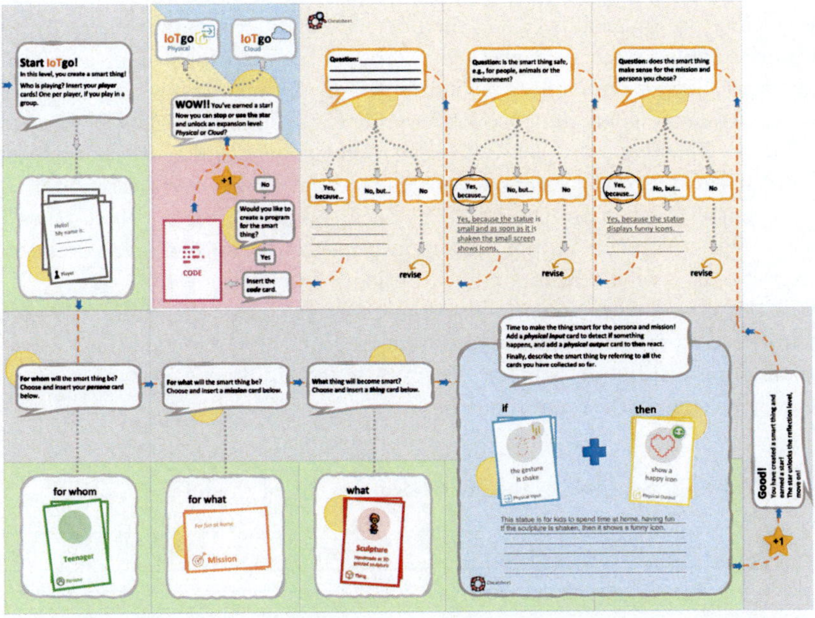

Fig. 1. A filled-in physical board of IoTgo

for smart things in the MakeCode environment, which follow and teach typical patterns for smart things interacting with people; (2) testing them rapidly with micro:bit physical-computing boards and devices [14]. The IoTgo app with MakeCode for micro:bit, English version, is partly accessible at https://share.streamlit.io/iotgo-app/iotgo-io/main/versions/bz_teachers_EN.py, protected by a non-commercial, share alike CC license.

Over time, the IoTgo toolkit has been co-designed with diverse people, such as pupils of different school levels, university students of art and design and of applied linguistics, besides professional artists, e.g., see https://made4me.it/iotgoarts/ and [5,7,17]. Usages and co-design actions led to the evolution of IoTgo, making it more modular, adaptable and adaptive, catering to varying needs, desiderata and expertise. The version presented in this paper was specifically adapted to school teachers.

3.4 IoTgo Protocol

Table 2 recaps the schedule of tasks of the intervention for teachers, organised per day. Each day had shared as well as specific tasks with IoTgo for moving teachers, tangibly, from design thinking into computational thinking, and evolve a smart-thing solution for an initially wicked problem.

Each day, specific tasks for design or computational thinking were gamified with IoTgo, so as to be replicable as-is with learners at school. For instance, during Day 2, participants had the following task: to play a conceptualisation game with IoTgo for groups of 3–4 members. Firstly, groups were split in pairs, whenever feasible. Each pair was given "rough" descriptions of ideas of smart things for tackling a problematic situation for given personas, with missions representing personas' goals in certain environments. These ideas had been created by children. Teachers were asked to revise and conceptualise ideas with input and output cards of IoTgo so as to abstract away details and decompose them with the pattern given by the IoTgo physical board in Fig. 1. Next pairs had to share their results in groups and reflect in groups on challenges related to children's ideas, which teachers may experience in class as well.

Each day ended with a common task, namely, to share reflections all together. Therein, researchers and participants reflected on what teachers had learnt in terms of design thinking and/or computational thinking, what was clear or unclear, what the next steps would be.

Table 2. Schedule of the main tasks per day of the intervention.

When	What for
Day 1	Exploring, reflecting
Day 2	Exploring, ideating, conceptualising, programming, reflecting
Day 3	Ideating, conceptualising, programming, prototyping, generalising, reflecting
Day 4	Programming, prototyping, generalising, reflecting

4 Study Results

The following part reports results of data collected through the pre-post questionnaire, administered in Day 1 and 4. It reports results concerning firstly the closed-format part of the questionnaire in Table 1, and secondly the open-format part asking to report freely ideas of design and computational thinking.

4.1 Closed Format

Teachers had to assess 3 groups of statements (Q1–Q3) on a 5-point Likert scale:

Q1. Computational/design thinking is useful in technical-scientific subjects.
Q2. Computational/design thinking is fundamental in today's society.
Q3. Computational/design thinking is useful in humanistic and artistic-musical.

Data were kept only if teachers provided answers to the pre- and post-questionnaire (18). For analysing answers, "absolutely no" was coded as −2, "no" as −1, "neutral" as 0, "yes" as 1, "absolutely yes" as 2. Overall, means for answers by teachers tended to increase after the intervention: from a 0 representing neutrality or uncertainty, especially for design thinking, towards 2 for

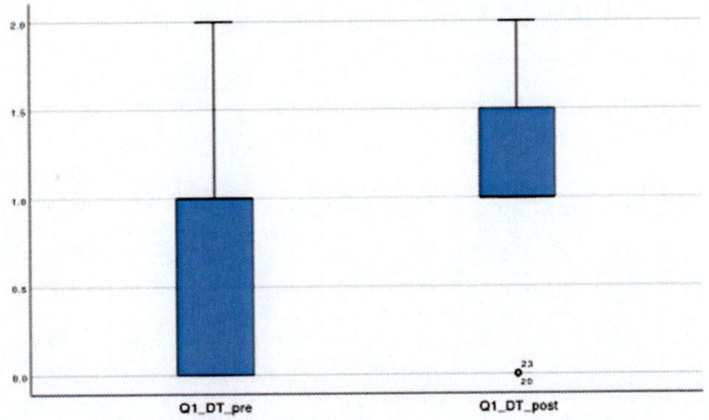

Fig. 2. Box-plot for the Q1-statement for design thinking, pre and post intervention

Table 3. Pre and post Means (M) and standard deviations (SD) computed with SPSS

Pre item	M	SD	Post item	M	SD
Q1_CT_pre	1.2400	.43589	Q1_CT_post	1.3200	.47619
Q1_DT_pre	.8750	.67967	Q1_DT_post	1.1667	.56466
Q2_CT_pre	1.2083	.50898	Q2_CT_post	1.3333	.56466
Q2_DT_pre	.708333	.750604	Q2_DT_post	1.083333	.653863
Q3_CT_pre	.833333	.564660	Q3_CT_post	1.041667	.550033
Q3_DT_pre	.666667	.761387	Q3_DT_post	1.041667	.690253

"absolutely yes". See Table 3 for means and standard deviations, and Fig. 2 for the box-plot related to the Q1-statement for design thinking.

We used a paired t test to test if there was a significant difference in the pre- and post-means. As Table 4 shows, with the exception of the pre- and post-means for the Q3 statement for computational thinking, for all other statements there was a statistically significant average increase following the intervention.

4.2 Open Format

Two researchers performed a deductive thematic analysis of teachers' answers to the two open-format questions concerning their ideas of computational thinking and design thinking, respectively: they looked for terms similar to those pertaining to design and computational thinking from the literature. They worked first independently and then they compared their analysis. Data were kept only if teachers provided answers to both questions (19). Thus, researchers counted such terms. Terms for computational thinking which were counted were: pattern, abstraction, problem-decomposition, algorithm, program. For design thinking they were: empathy, problem/situation, idea, prototype. The relative frequencies were computed for computational-thinking terms, pre and post (FCT_pre,

Table 4. Paired t test results with SPSS

Item	N	Correlation	Sig.
Q1_CT_pre & Q1_CT_post	25	.618	**.001**
Q1_DT_pre & Q1_DT_post	24	.623	**.001**
Q2_CT_pre & Q2_CT_post	24	.504	**.012**
Q2_DT_pre & Q2_DT_post	24	.760	**.000**
Q3_CT_pre & Q3_CT_post	24	.303	.150
Q3_DT_pre & Q3_DT_post	24	.689	**.000**

FCT_post), and design-thinking terms, pre and post (FDT_pre, FDT_post). In all cases, there was an increase in the mean relative frequency of terms related to design and computational thinking, following the intervention: for FCT_pre, the mean is 0.283, standard deviation is 0.235, and for FCT_post, the mean is 0.408, standard deviation is 0.253; for FDT_pre, the mean is 0.219, standard deviation is 0.185, and for FDT_post the mean is 0.333, standard deviation is 0.262.

We run the paired t test on the relative frequencies of terms associated to computational thinking and to design thinking, before and after the intervention. According to the outcome of the test, the intervention seems to have elicited a statistically significant increase in the mean relative frequency of terms associated to computational thinking ($t(23) = -2.128$, $p = .044$) and in that of terms associated to design thinking ($t(23) = -2.2$, $p = .038$).

5 Discussion and Conclusions

This paper outlines an intervention for 32 Italian primary and secondary school teachers concerning design and computational thinking. The work intercepts recent recommendations for interventions guiding teachers to both design and computational thinking, and leveraging on their understanding of them [11,19]. This paper reports results of a pre-post questionnaire with closed-format and open-format questions, investigating teachers' ideas of both types of thinking. They are discussed in the remainder, together with limitations and future work.

5.1 Discussion of the Study Results

Computational and design thinking are both considered relevant skills for learners. In spite of that, as the background section shows, they are seldom taught together. One of the many barriers are teachers' ideas of them, besides "a need for additional guidance, case studies, and other forms of teacher professional development" [19]. The intervention reported in this paper was organised tangibly in hands-on laboratories with IoTgo, so as to be replicable at school as-is. Results of the data analyses seem to point out the effectiveness of the intervention.

Both the closed- and open-format part of the questionnaire show a change in teachers' ideas. The closed-format part asked teachers to assess three groups of statements (Q1–Q3), related to computational and design thinking; see Table 1. Given that the large majority of participants (96%) teach maths, science or technology subjects, their answers to Q1 and Q2 are particularly relevant. According to the result of a t-test analysis, these statements received higher points after the intervention, indicating that such teachers tended to perceive both of them more useful for their subjects and society, following the intervention.

The open-format part of the questionnaire asked teachers to freely report their ideas of computational and design thinking. Their ideas were analysed, counting terms which are found in definitions of the two forms of thinking, presented the first day of the intervention. Their relative frequencies tended to increase after the intervention, and statistically significantly so. This is taken as an indication that teachers had internalised the concepts they had mastered during the intervention via hands-on activities, guided by the IoTgo toolkit.

5.2 Limitations and Future Work

The main limitation of the reported study is that it only analyses teachers' ideas, before and after the intervention, although in two different manners. During the intervention, teachers also produced their own smart things starting from wicked problems, by means of the IoTgo toolkit, and reflected over them. Future work will analyse teachers' artefacts and reflections so as to complement the findings reported in this paper, and study what to adapt of the IoTgo toolkit to best match teachers' mental models [4,6]. Moreover, several participants already implemented what they had learnt with their own classes. Their activities are still on-going at the time of writing. Future work will consider what they have done in class as a further indicator of the effectiveness of the intervention reported in this paper.

References

1. Chamberlain, L., Mendoza, S.: Design thinking as research pedagogy for undergraduates: project-based learning with impact. Counc. Undergraduate Res. Q. **37**, 18–22 (2017). https://doi.org/10.18833/curq/37/4/15
2. Corradini, I., Lodi, M., Nardelli, E.: Conceptions and misconceptions about computational thinking among Italian primary school teachers. In: Proceedings of the 2017 ACM Conference on International Computing Education Research, ICER 2017, pp. 136–144. Association for Computing Machinery, New York (2017). https://doi.org/10.1145/3105726.3106194
3. Crichton, H., McDaid, A.: Learning intentions and success criteria: learners' and teachers' views. Curric. J. **27**(2), 190–203 (2016). https://doi.org/10.1080/09585176.2015.1103278
4. Di Mascio, T., Gennari, R., Melonio, A., Tarantino, L.: Supporting children in mastering temporal relations of stories: the TERENCE learning approach. Int. J. Distance Educ. Technol. **14**(1), 44–63 (2016). https://doi.org/10.4018/IJDET.2016010103. Cited by: 11

5. Gennari, R., Melonio, A., Rivi, M., Matera, M.: Physical or on the cloud: play with IoTgo and design smart things. In: CEUR Proceedings of the Joint Workshop on Games-Human Interaction (GHItaly 2021) and Multi-Party Interaction in eXtended Reality (MIXR 2021) Co-Located with CHItaly 2021: 14th Biannual Conference of the Italian SIGCHI Chapter, Bolzano, Italy, 12 July 2021 (2021)
6. Gennari, R., Melonio, A., Rizvi, M.: Evolving tangibles for children's social learning through conversations: beyond turntalk. In: Proceedings of the Twelfth International Conference on Tangible, Embedded, and Embodied Interaction, TEI 2018, pp. 368–375. Association for Computing Machinery, New York (2018). https://doi.org/10.1145/3173225.3173248
7. Gennari, R., Rizvi, M.: At the frontiers of art and IoT: the IoTgo toolkit as a probe for artists. In: CHItaly 2021: 14th Biannual Conference of the Italian SIGCHI Chapter, CHItaly 2021. Association for Computing Machinery, New York (2021). https://doi.org/10.1145/3464385.3464727
8. Gero, J.S.: Design prototypes: a knowledge representation schema for design. AI Mag. **11**(4), 26 (1990)
9. Plattner, H., Meinel, C., Leifer, L.: Design Thinking: Making Design Thinking Foundational. Springer, Cham (2016)
10. Hennessey, E., Mueller, J.: Teaching and learning design thinking (DT): how do educators see DT fitting into the classroom? Can. J. Educ./Revue canadienne de l'éducation **43**(2), 498–521 (2020)
11. Kelly, N., Gero, J.S.: Design thinking and computational thinking: a dual process model for addressing design problems. Des. Sci. **7**, e8 (2021). https://doi.org/10.1017/dsj.2021.7
12. Margot, K.C., Kettler, T.: Teachers' perception of stem integration and education: a systematic literature review. Int. J. STEM Educ. **6**, 1–16 (2019). https://doi.org/10.1186/s40594-018-0151-2
13. Melles, G., Howard, Z., Thompson-Whiteside, S.: Teaching design thinking: expanding horizons in design education. Procedia. Soc. Behav. Sci. **31**, 162–166 (2012). https://doi.org/10.1016/j.sbspro.2011.12.035
14. BBC micro:bit (2022). https://microbit.org/
15. Razzouk, R., Shute, V.: What is design thinking and why is it important? Rev. Educ. Res. **82**, 330–348 (2012). https://doi.org/10.3102/0034654312457429
16. Retna, K.S.: Thinking about "design thinking": a study of teacher experiences. Asia Pac. J. Educ. **36**(sup1), 5–19 (2016). https://doi.org/10.1080/02188791.2015.1005049
17. Rizvi, M.: Supporting end users in designing IoT smartthings with the IoTgo toolkit. In: CEUR Proceedings of the 2nd International Workshop on Empowering People in Dealing with Internet of Things Ecosystems (EMPATHY) Co-Located with INTERACT 2021, Bari, Italy (2021)
18. Rotherham, A.J., Willingham, D.T.: 21st century skills: the challenges ahead. Educ. Leadersh. **67**, 16–21 (2009)
19. Slotta, J.D., Chao, J., Tissenbaum, M.: Fostering computational thinking and design thinking in the PYP, MYP and DP (2020)
20. Wing, J.: Computational thinking. Commun. ACM **49**, 33–35 (2006). https://doi.org/10.1145/1118178.1118215

Brewing Umqombothi: Technicalities of a VR Prototype Merging STEM and South African Intangible Cultural Heritage

Kasper Rodil[1](✉), Mihai Ciungu[1], Peter Leth[1], Steffan Christensen[1], Umesh Ramnarain[2], and Mafor Penn[2]

[1] Department of Architecture, Design and Media Technology, Aalborg University, Rendsburggade 14, 9000 Aalborg, Denmark
kr@create.aau.dk, {mciung18,pleth18,schri18}@student.aau.dk
[2] Department of Science and Technology Education, Faculty of Education, University of Johannesburg, B Ring 436, Kingsway Avenue, Auckland Park, Johannesburg 2006, South Africa
{uramnarain,mpenn}@uj.ac.za

Abstract. This paper describes an interactive embodied learning-based simulation in Virtual Reality that has been designed to disseminate the Intangible Cultural Heritage behind Umqombothi, a traditional South African beer-brewing practice, by instructing both the cultural process and underlying chemistry. The VR simulation includes an authentic beer-making environment based on a South African township and an abstract chemistry environment called the "Microverse". The paper is limited to literature grounding and details of the technical prototype game and its implicit logging system.

Keywords: Intangible cultural heritage · Craftsmanship · Embodied interaction · STEM · Virtual reality

1 Introduction

UNESCO [1] defines Cultural Heritage as not only physical artifacts such as monuments or collections or objects, but also traditions and ideas passed through generations, known as Intangible Cultural Heritage (ICH). ICH can be oral traditions, social practices, or traditional crafts. The ICH is important mainly through its extensive wealth of knowledge and skills, which have great socioeconomic benefits in minority groups [1]. An example of ICH is Umqombothi, a traditional South African beer, rich in calories, B vitamins and essential amino acids [2]. The beer is the epitome of ICH, as it has a vast socioeconomic importance while being a tradition passed down orally through generations. This paper suggests a solution for safeguarding South African ICH through embodied learning within a Virtual Reality simulation. The simulation aims to enable learning of the fundamentals of beer-making, as a cultural process, and the chemistry involved in making Umqombothi following the instructions by Hlangwani et al. [3].

M. Temperini et al. (Eds.): MIS4TEL 2022, LNNS 580, pp. 175–180, 2023.
https://doi.org/10.1007/978-3-031-20617-7_22

The work presented here limits to the technical prototype from a larger project on how to seek opportunities for bringing the ICH of rural community groups in South Africa into the STEM curriculum of the Primary School using VR.

The project has a nucleus of collaborating closely with the communities in articulating ICH that might be relevant to learn from a cultural, identity perspective for young learners to see contemporary value in their cultural background. On the other hand, the interest in STEM related parts of the curriculum among young learners is not overwhelming, and continues to be a focus point in national, educational strategies [4]. Jiang et al. [5] suggest that early exposure to technologically enhanced virtual environments could spark and maintain a long-term interest of students which could lead to pursuing careers in STEM fields.

Thus, the aim is bi-focal, ensuring cultural cohesion meanwhile invigorating the STEM curriculum.

2 Related Work

VR has greatly increased in popularity in the last few years [6], and institutions such as museums and formal education are reorganizing themselves to enable new learning opportunities meanwhile seeking to be relevant to their audiences. Whereas youths in local communities are fascinated by cultural content, be it games or streaming media, usually not being provided by their own communities.

The digitization of culture is not a new idea. Zara [7] presented techniques for interacting with cultural heritage objects on the web back in 2004. But when it comes to VR much work has been devoted to the tangible aspects of cultural heritage. For example, Fassi et al. [8] used VR for cultural heritage as support for conservation and maintenance activities for the Milan Cathedral. Another example is the digitization of Canadian cultural objects, demonstrating how the use of VR motivated students and its usefulness in educational settings [9].

Scoping VR as a technological opportunity to disseminate ICH has not been fully explored yet, although related work exists. For instance, protection and dissemination of traditional craftsmanship practices, such as glassblowing [10] or carpentry [11].

One of the challenges is the interpretation of expressions, practices and beliefs that are abundant among living communities, and the challenge of transferring from one modality (e.g., written sources) into multi-modal VR systems [12].

Ch'ng et al. [13] argue that VR technology is one of the most promising technologies for facilitating the learning of culture and heritage, for example through integration in museums. Seeing ICH craftsmanship and practices as being highly performative by nature requires two connected components; how can the user instead of being a passive receiver of dissemination, instead interact [14] with the content?

2.1 Learning Through Embodied Interaction

Recent research proposes VR as a technology with a tremendous potential in digitalizing contemporary culture, as it has the possibility of embodied interaction with a non-symbolic approach to computing systems, allowing non-technical individuals to participate in the interaction and dissemination of cultural content [15].

Cunha et al. [16] aimed to find what facilitates learning from the students' perspective. They have found two major "dimensions" in their study: the teacher and the student, with the teacher being the most visible in their data. Their findings show that students reported a high correlation between the learning process and characteristics of the teacher (e.g., way of teaching, instructional clarity, personality), as well as the teacher's ability to make the content engaging and fun. The same study shows that "activities that lead to action and the active participation of students" were appreciated by the participants. Research has shown that embodied interaction have been proven to be very effective in STEM areas [17]. An embodied interaction, in an educational context, means that the learner initiates a physical movement that is mapped to the content to be learned [18]. Lindgren et al. [17] also suggest that embodied interaction is likely to result in more engagement from students than classic learning methods. However, there are currently no design instructional guidelines for VR simulations that rely on embodiment [18], which mostly speaks to the highly varied content of the subject and the multitude of ways one can create VR systems.

By relating Cunha [16] and Lindgren's [17] findings, there is apparent potential in replacing the human teacher with gamified computer simulations that aim to greatly increase the engagement of the student in the learning process - also in the context of disseminating cultural content.

3 The Development of a Virtual Reality Prototype

Fig. 1. The figure displays the environments in the simulation. Left the tutorial scene, middle the township scene where the user brews beer. Right shows the Microverse for interactions with the chemical processes.

The simulation (see Fig. 1) was built in Unity and the VR functionality was facilitated by SteamVR, allowing for the simulation to be deployed on a wide range of VR headsets. In the simulation, the player moves through teleportation. A comprehensive event-based interaction system was implemented specifically for facilitating the interactions in the simulation. The beer-making and the chemistry had two different teaching environments that the player would switch between, with an additional environment acting like a short tutorial and an introduction to the narrative of the simulation. All the 3D models in the environments were made in Autodesk Maya 2022.

3.1 Gameplay Overview and Logging System

In the simulation, the goal of the player is to take the necessary beer-making steps while also understanding the underlying chemistry interactions.

The beer-making is facilitated by traditional South African tools within the realistic township environment shown in Fig. 1. The chemistry environment is abstract, with the player being tasked with shooting molecules inside the Microverse to trigger the chemical reactions that happen during the fermentation process, using an "enzyme gun". The gun itself further helps the player understand the enzymes involved in the chemical process, by displaying their names on a small screen, while also offering hints on the progress of the player towards completing the interaction.

The objects that the player can interact with are highlighted to signify affordance. Furthermore, a diegetic voiceover is present throughout the simulation, using an authentic South African voice. The purpose of the voiceover is twofold: presenting a simple narrative and explaining the interactions in the simulation. Establishing voicelines played before every first interaction in the township and in the Microverse. In some cases, further voicelines were added before certain actions in the township for added context. In the Microverse, after one shot is fired in each level, an additional explanatory voiceline plays. As such, the teaching in the township environment is embodied to increase engagement, as per Cunha [16] and Lindgren [17]. The instruction of the beer-making is, however, enhanced through verbal instruction through the diegetic voiceover. The teaching in the chemistry Microverse leans more towards being verbal and visual, as the mapping of the chemical processes to embodied interaction are by nature too abstract without supporting instruction.

A logging system is embedded in the simulation to gather evidence about the proper functionality of the VR solution, e.g. frames per second (FPS), and can be used to analyze user behaviors during their VR experiences [19, 20], potentially providing a new perspective on evaluation results [21]. Implicit logging has the additional benefit of being a non-intrusive way of capturing significant amounts of data [22]. An example of data captured, and a use case is shown on Fig. 2.

Combined FPS logging and player position logging can provide insights into time spent in instructional VR context, highlighting how the township scene sees significantly lower FPS than the other scenes in the simulation. Visible by the heatmap graph on the bottom suggests that the township, while costly to render, is also not fully utilized. This serves as an example, of many, how implicit data capture can aid in designing better interactive experiences in VR.

4 Future Work

Although promising preliminary results from the initial evaluation have been reported, they are beyond the scope of this paper. Further refinement of the technical prototype and future experimentation are planned.

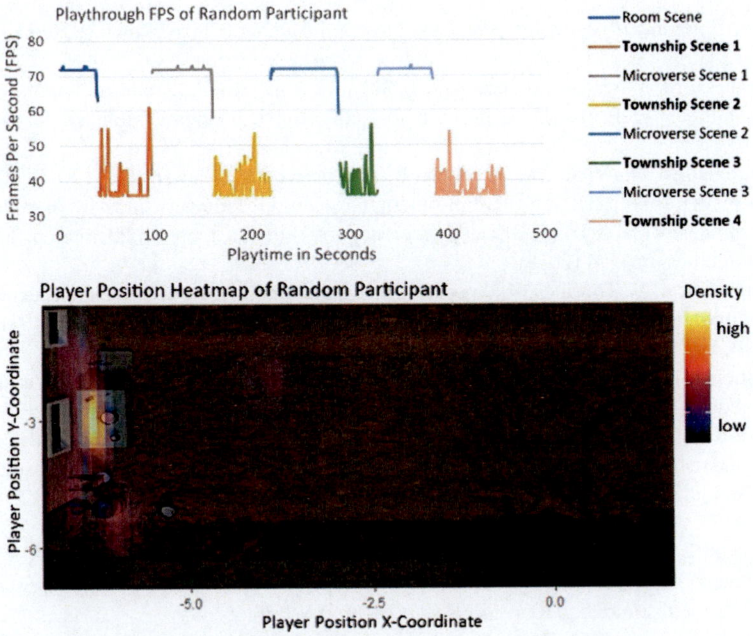

Fig. 2. The figure shows FPS logs (top) and corresponding visualization of player positions within the township scene (bottom)

References

1. UNESCO: What is Intangible Cultural Heritage? - Intangible Heritage - Culture Sector - UNESCO. https://ich.unesco.org/en/what-is-intangible-heritage-00003
2. Lyumugabe, F., Gros, J., Nzungize, J., Bajyana, E., Thonart, P.: Characteristics of African traditional beers brewed with sorghum malt: a review (2012)
3. Hlangwani, E., Adebiyi, J.A., Doorsamy, W., Adebo, O.A.: Processing, characteristics and composition of umqombothi (a South African traditional beer). Processes **8**(11), 1451 (2020). https://doi.org/10.3390/pr8111451
4. Sadler, P.M., Sonnert, G., Hazari, Z., Tai, R.: Stability and volatility of STEM career interest in high school: a gender study. Sci. Educ. **96**(3), 411–427 (2012). https://doi.org/10.1002/SCE.21007
5. Jiang, Y., Popov, V., Li, Y., Myers, P.L., Dalrymple, O., Spencer, J.A.: "It's like i'm really there": using VR experiences for STEM career development. J. Sci. Educ. Technol. **30**(6), 877–888 (2021). https://doi.org/10.1007/s10956-021-09926-z
6. Stevanovic: 30 Virtual Reality Statistics for 2020. KommandoTech, December 2019. https://kommandotech.com/statistics/virtual-reality-statistics/
7. Zara, J.: Virtual reality and cultural heritage on the web. In: Proceedings of the 7th, June 2004. https://www.academia.edu/21282758/Virtual_reality_and_cultural_heritage_on_the_web
8. Fassi, F., Mandelli, A., Teruggi, S., Rechichi, F., Fiorillo, F., Achille, C.: VR for cultural heritage. In: De Paolis, L.T., Mongelli, A. (eds.) AVR 2016. LNCS, vol. 9769, pp. 139–157. Springer, Cham (2016). https://doi.org/10.1007/978-3-319-40651-0_12
9. Garcia-Ruiz, M.A., Santana-Mancilla, P.C., Gaytan-Lugo, L.S.: A user study of virtual reality for visualizing digitized Canadian cultural objects. In: Cases on Immersive Virtual Reality Techniques, pp. 42–66. IGI Global (2019)

10. Carre, L., et al.: Mixed-reality demonstration and training of glassblowing. Heritage **5**(1), 103–128 (2022)
11. Rossau, G., Skovfoged, M.M., Czapla, J.J., Sokolov, M.K., Rodil, K.: Dovetailing: safeguarding traditional craftsmanship using virtual, reality. Int. J. Intangible Heritage **14**, 103–120 (2019)
12. Skovfoged, M.M., Viktor, M., Sokolov, M.K., Hansen, A., Nielsen, H.H., Rodil, K.: The tales of the Tokoloshe: safeguarding intangible cultural heritage using virtual reality. In: Proceedings of the Second African Conference for Human Computer Interaction: Thriving Communities, pp. 1–4 (2018)
13. Ch'ng, E., Cai, Y., Thwaites, H.: Special issue on VR for culture and heritage: the experience of cultural heritage with virtual reality: guest editors' introduction. Presence **26**(03), iii–vi (2018). https://doi.org/10.1162/pres_e_00302
14. Schofield, G., et al.: Viking VR: designing a virtual reality experience for a museum. In: Proceedings of the 2018 Designing Interactive Systems Conference, pp. 805–815 (2018)
15. Rodil, K., Maasz, D., Winschiers-Theophilus, H.: Moving virtual reality out of its comfort zone and into the African Kalahari desert field: experiences from technological co-exploration with an Indigenous San community in Namibia, June 2020. https://doi.org/10.1145/3385956.3418955
16. Cunha, R.S., Ribeiro, L.M., Sequeira, C., de Almeida Barros, R., Cabral, L., Dias, T.S.: What makes learning easier and more difficult? The perspective of teenagers. Psicologia em Estudo **25** (2020). https://doi.org/10.4025/PSICOLESTUD.V25I0.46414
17. Lindgren, R., Tscholl, M., Wang, S., Johnson, E.: Enhancing learning and engagement through embodied interaction within a mixed reality simulation. Comput. Educ. **95**, 174–187 (2016). https://doi.org/10.1016/J.COMPEDU.2016.01.001
18. Johnson-Glenberg, M.: Immersive VR and education: embodied design principles that include gesture and hand controls. Front. Robot. AI **5** (2018). https://doi.org/10.3389/frobt.2018.00081
19. Luoto, A.: Systematic literature review on user logging in virtual reality. In: ACM International Conference Proceeding Series, pp. 110–117, June 2018. https://doi.org/10.1145/3275116.3275123
20. Steptoe, W., Steed, A.: Multimodal data capture and analysis of interaction in immersive collaborative virtual environments (2012)
21. Ritchie, J.M., Dewar, R.G., Robinson, G., Simmons, J.E.L., Ng, F.M.: The role of non-intrusive operator logging to support the analysis and generation of product engineering data using immersive VR. Virtual Phys. Prototyping **1**(2), 117–134 (2006). https://doi.org/10.1080/17452750600763947
22. Ritchie, J.M., Sung, R.C.W., Rea, H., Lim, T., Corney, J.R., Howley, I.: The use of non-intrusive user logging to capture engineering rationale, knowledge and intent during the product life cycle. In: PICMET: Portland International Center for Management of Engineering and Technology, Proceedings, pp. 981–989 (2008). https://doi.org/10.1109/PICMET.2008.4599707

Serious Games for Autism Based on Immersive Virtual Reality: A Lens on Methodological and Technological Challenges

Vita Santa Barletta[1], Federica Caruso[2](✉), Tania Di Mascio[2], and Antonio Piccinno[1]

[1] University of Bari Aldo Moro, 70121 Bari, Italy
{vita.barletta,antonio.piccinno}@uniba.it
[2] University of L'Aquila, 67100 L'Aquila, Italy
federica.caruso1@graduate.univaq.it, tania.dimascio@univaq.it

Abstract. The use of Serious Games in the treatment of people with Autism Spectrum Disorders is nowadays considered promising, given the positive effects in promoting the acquisition of learning through motivating and engaging experiences. In particular, recent years have seen increased research attention toward serious games based on Immersive Virtual Reality technologies (i.e., large-scale projection-based systems, head-mounted displays). This is due to the benefits that the high level of immersion produces in learning outcomes: a high level of immersion eliminates environmental distractions and fosters attention on the learning tasks as well as coping with anxiety and social phobias. Unfortunately, the design and evaluation of these interventions are not without challenges and issues, given the difficulties in applying adequate and rigorous methodological design approaches. This paper presents a review of the available Serious Games for Autism based on Immersive Virtual Reality, developed between 2009 and mid-2021, selected and categorized with respect to target users' characteristics, learning purpose, and the adopted Immersive Virtual Reality technologies. The analysis mainly put a lens on technological aspects and on how these interventions were designed, as well as on the people involved in the design itself. Finally, some preliminary guidelines are delineated.

Keywords: Autism Spectrum Disorders · Serious game · Immersive Virtual Reality · Systematic literature review

1 Introduction

The growing diffusion of Information and Communication Technology (ICT) and the concomitant system cost reductions encouraged the exploitation of innovative tools in the education, therapy, and habilitation/rehabilitation of several clinical populations, especially the Autism Spectrum Disorders (ASD) [6,10].

M. Temperini et al. (Eds.): MIS4TEL 2022, LNNS 580, pp. 181–195, 2023.
https://doi.org/10.1007/978-3-031-20617-7_23

The importance of therapeutically intervening in people with ASD has never been greater: in the last decade, the prevalence of autism has increased (recent statistic estimating that 1% of the world's population has been identified with ASD [32,44]). As a result, since the 1970s, a great deal of research seeks to identify effective ICT interventions to improve social, practical, and conceptual skills of individuals with ASD [9,10,54].

The analysis of the available literature suggests that Serious Games (SGs) are the most promising learning approach for therapeutic interventions for ASD conditions, notably if they target social skills [54,57]. In fact, several studies reported positive outcomes in terms of acquired knowledge and demonstrate that SGs could promote (1) a motivating and engaging acquisition of learning and (2) generalization and transfer of skills learned by playing in real life [17,27,53,54,57,59]. As defined by Clark Abt [2], Serious Games are games that "have an explicit and carefully thought-out educational purpose and are not intended to be played primarily for amusement. This does not mean that serious games are not, or should not be, entertaining". As a result, designing SGs entails the merge of learning theory and empirical findings together with the principles of game design to create a unique intervention tool that can target any set of skills (e.g., cognitive, behavioural, and social skills) to improve knowledge and competencies beyond the context of the game. Given the very nature of SG, only a multidisciplinary team composed of different expertise can produce games well balanced between learning and entertainment aspects [35]; this is particularly crucial when the SG target audience presents specific characteristics, needs, and attitudes such as people with ASD [54].

Reviewing existing SGs for individuals with ASD, it emerged that they were increasingly implemented as Virtual Reality (VR) applications with positive learning outcomes given the high visual processing skills possessed by people with ASD [4,17,53,54]. *Virtual Reality* refers to a technology based on computer graphics capable of creating virtual scenes and objects that can be manipulated by the user through input devices and that can be seen, heard, touched, or even smelt through output devices [60]. Consequently, differently from other computer technologies, the users can feel immersed within the virtual environment as if they are "really" there [15]. As asserted by [15,38], the level of immersion delivered by technological means plays a key role, since it can influence the learning of people with ASD and their involvement. In particular, it emerged that the optimal level of immersion for this clinical population is that provided by VR technologies offering a moderate/high level of immersion [38], i.e., Immersive Virtual Reality (IVR) technologies. The main IVR technologies are large-scale projection-based systems, such as Cave Automatic Virtual Environments (CAVEs) and Head-Mounted Displays (HMDs) [15]. There is a great deal of research work demonstrating the effectiveness of IVR as an intervention tool for people with ASD, especially for learning and training social skills [15,33,37,38]. The immersion can enhance the learning outcomes by eliminating environmental distractions and fostering the attention on the learning task to perform [15]. Furthermore, these technologies can improve the ability of people with ASD to deal with anxiety and phobias in social situations [15].

Given these premises, the merge of game-based learning experience with IVR technologies enables a new generation of more effective ICT-based therapeutic tools in ASD field: IVR-based SG for the treatment of people with ASD. However, the design of these interventions is not without challenges and issues, given the lack of methodological approaches to follow [18,30,53]. In particular, as shown in a previous study [18], there is a lack of design methodologies that could be adopted "*as-is*" in the design of IVR-based SG targeted to individuals with ASD. From a literature investigation, it emerged that there are only general guidelines on game elements proving to have a positive impact on the learning of people with ASD [57] or frameworks including a set of game elements that could improve the learning of some specific skills by children with ASD [30,42]. Including even SG design frameworks for typical users, the available literature lacks some essential aspects that, rather than others, play a key role in designing effective, engaging, and accessible IVR-based SGs for people with ASD. For example, the characterization of target users [17,54], the technology to be adopted as output devices [54], the methodology to follow within the design process [35], the composition of the multidisciplinary design team [35], and the active participation of representatives of the target users [53].

Therefore, to take a step toward solving this lack, it would be valuable to investigate previous IVR-based SGs targeting people with ASD, to identify recent trends and common methodological approaches meaningful to propose an innovative IVR-based SG design methodology that encompasses all the above-mentioned challenges.

The aim of this paper is to provide a review of available IVR-based SGs with respect to characteristics of users' target, learning purpose, and the IVR technologies adopted, to identify recent trends in this field. Furthermore, an analysis of how these interventions were designed is presented, focusing on the design methodology adopted and the composition of the team involved in the design itself, with the main objective of finding common design indications for IVR-based SGs in the literature.

The rest of the paper is organized as follows. Section 2 presents an overview of the existing literature review on SG and IVR in the ASD domain, revealing the absence of a study focusing on the joint adoption of SG and IVR. Section 3 explains the research methodology adopted and data extraction method, along with the inclusion/exclusion criteria. In Sect. 4, the results of this review are presented and discussed. Finally, Sect. 5 presents the conclusions drawn after the analysis of the studies and introduces future research activities.

2 Related Work

This review aims to identify and analyse the available IVR-based SGs for people with ASD to delineate some preliminary guidelines. To complete this task, existing literature reviews in this field were studied and analysed. Unfortunately, to the best of our efforts, the investigation of several digital libraries (e.g., Scopus, Google Scholar, ACM digital library, Science Direct) did not reveal the existence of a review focused on SGs for people with ASD implemented as IVR applications. Generally, the discovered studies focused on SG for ASD, or vice versa on IVR for ASD, not on their joint adoption. Regarding existing reviews on SGs or, more generally, on gamification aspects of interventions for people with ASD, there are several reviews (e.g., [17,27,28,53,54]). However, they generally focus their analysis on SGs presenting a specific characteristic, such as target learning skills (e.g., social skills [27]), target audience (e.g., children with ASD [28,59], and technology means (e.g., computer-based interventions [53]). Among these reviews, the work proposed by Tsikinas and Xinogalos [2018] [53] is especially interesting, as one of the research questions addressed in this review was the design methodologies adopted for the development of SG for people with intellectual disabilities and people with ASD. According to the analysis of the 54 SGs selected in this review, few are the works claiming the design methodology followed (three related to SG for intellectual disabilities and four to SG for ASD). Specifically, the design methodologies adopted are Participatory Design [24], User-Centered Design [1], and Learner-Centered Design [49]. Unfortunately, none of the SGs analyzed in this work is implemented as an IVR application. On the other hand, regarding existing reviews on IVR for ASD, their primary aim is to describe the current state-of-the-art [26,33,37] and identify shared immersive virtual environment design choices, without reference to the learning approach eventually adopted [15].

Thus, a unique feature of this review in comparison with existing literature reviews is that it analyses the available IVR-based SGs for people with ASD, in an attempt to provide a comprehensive analysis focused on their methodological and technological aspects; the outcome of this analysis will delineate some preliminary design guidelines. It is our belief that researchers can benefit from this analysis.

3 Research Methodology

A systematic literature review was conducted following the guidelines shared by Kitchenham [31].

To identify studies to be included in this review, a computerized search of the following digital research databases was performed: (1) *Scopus*, (2) *ACM digital library*, (3) *IEEE Xplore Digital Library*, (4) *Science Direct*, (5) *Web of Science*, (6) *Semantic Scholar*, (7) *PubMed*, (8) *Google Scholar*. In all digital research databases, the search was limited to papers published from 2009 to July 2021, thus capturing recent studies adopting the current generation of IVR technology. The following criteria were used for the search:

- **Search String** Autism AND Immersive Virtual Reality AND Serious Game
- **Synonyms/Abbreviations** ASD, IVR, Educational Game
- **Inclusion Criteria** • Studies published between January 2009 and July 2021 • Studies written in English • Books, chapters, journal and proceedings papers • Studies with focus on ASD • Game-based intervention tools with

an educative or training purpose • Studies including design and evaluation process • Full Paper available (not only abstract).
- **Exclusion Criteria** • Studies published before 2009 • Studies not written in English (e.g., Chinese, Portuguese) • Contributions of the following types: reviews, letters, extended abstracts, editorials, notes, dissertations, theses, doctoral consortium, letters to the editor • Studies focusing on other mental illnesses (e.g., dementia and cerebral palsy) • Applications not matching the definition of Serious Game provided by Clark Abt [2] • Robotic solutions, physical games, or other applications not IVR-based • Studies with an exclusive medical focus or a focus on the diagnosis of ASD.

In addition, the bibliography of studies meeting the including criteria as well as the reviews resulting from the search were reviewed to identify supplemental studies for inclusion. Figure 1 presents the entire study selection process and the exact number of remaining papers in each step.

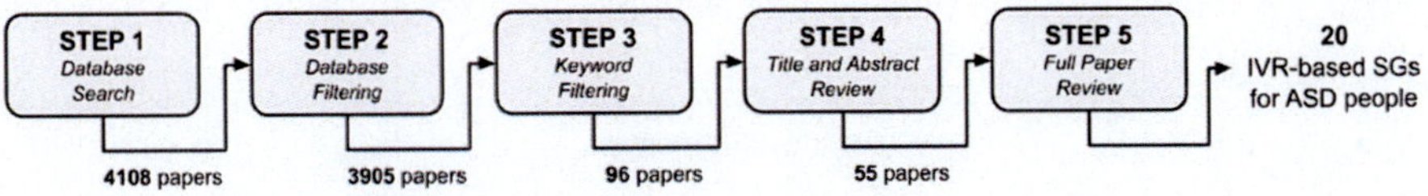

Fig. 1. Overview of the selection process.

First, a search with the formulated search string was done (Step 1) in each selected digital research database. Second, non-English contributions published before 2009, duplicates, and papers belonging to excluded categories (e.g., editorials, notes, dissertations, theses) were discarded (Step 2). Third, "keyword filtering" was carried out to discard any papers that did not mention in the body IVR technologies and serious games targeting ASD individuals (Step 3). Then, the title and abstract of the remaining papers were reviewed (Step 4). Finally, the entire body of papers was examined (Step 5). Please note that the papers discussing the same application were analyzed in aggregate (e.g., [11–14] refer to the same application, namely VR4VR, thus we extracted the relevant information by analyzing all of them).

Steps 1 to 3 were performed by a researcher with a background in ICT for ASD treatment. Steps of 4 to 5 were conducted by four researchers (i.e., two ICT-experts and two experts with a background in ASD and psychology), with any disagreement discussed to reach consensus.

After the study selection process, a series of relevant information was extracted from the retrieved studies. Specifically: the year of publication, the list of authors, target users' characteristics (i.e., the diagnosis and the age range), target learning skills (e.g., social, practical skill), the technology employed as output (e.g., CAVE, HMD), the design methodology (e.g., User-Centered Design, Participatory Design), the composition of the team involved in the design process (e.g., ASD-experts, ICT-experts). Table 1 resumes all information extracted from the selected IVR-based SGs targeting people with ASD.

4 Analysis and Discussion of the Results

Applying the selection criteria, a total of 20 studies were selected. As detailed above, only studies published between January 2009 and July 2021 were considered. In Fig. 2, the number of studies published in this time frame are plotted (in case of aggregated studies, we considered the year of the latest published). It is possible to notice an increase in publications on this topic over the period, confirming the growing popularity of game-based learning intervention tools [53]. Furthermore, this upward trend starts from 2013, in accordance with the launch of the first commercial HMD (i.e., the first Oculus Rift dk1). In addition, recent IVR technologies are increasingly promising than early released devices given their improvements, over the last years, in the quality of HMDs and their significant cost reduction, thus encouraging the exploitation of these innovative technologies [19]. Please, notice that the studies found in 2021 (i.e., [20,47]) are not reported in this plot because it would have been misleading to show incomplete data, as this review covered studies retrieved until July 2021.

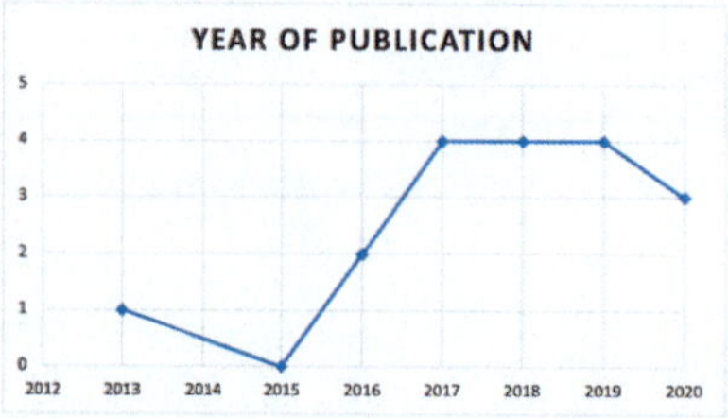

Fig. 2. Years of publication

Target Audience. The selected studies were analysed to identify the diagnosis and the age range of the target audience. It emerged that these therapeutic tools were mainly designed for individuals with ASD, except for two studies that targeted people with Neuro-Developmental Disorders (NDD) (including intellectual disability, attention deficit hyperactivity disorder, Down syndrome, and ASD) [25,56] and children with Social Impairments (SI) [22].

Focusing on the age range of the target audience, 80% of the included studies were designed for youth (0–19 years old), specifically 65% for children (13 studies) and 15% for adolescents (3 studies) (See Table 1 for more details). Instead, few studies reported an intervention targeting adults with High-Functioning ASD (HFA) (I.Q. $>$ 80) [12,46]. Finally, two studies did not specify the age range of the target audience of their proposals, although the participants in the evaluation sessions are young adults [48,56].

These findings are consistent with other published studies focused on SGs for ASD [27,53] and IVR-based interventions for ASD [15,26,37,38,43].

Table 1. Available IVR-based SGs for people with ASD

Ref.	Target user	Learning purpose	Technology	Design methodology	People involved
[3]	Adolescents with ASD	Practical skill (money management)	Tethered IVR headset (HCT Vive)	User-Centered Design	Special Educators
[5]	Children with ASD	Practic al skill (street-crossing)	Smartphone IVR headset (Xiaomi VR headset)	–	Teachers
[11–14]	Adults with HFA	Practical skill (work-related)	Tethered IVR headset (VR2200) +180° Curved curtain screen	–	Special Educators
[20]	Children with ASD	Cognitive skill (non-verbal fluid intelligence, attention processes, and sensorimotor integration skills)	Large-scale screen projections	–	–
[22]	Children with SI	Social skill (eye-contact)	Tethered IVR headset (HCT Vive)	Participatory Design	ASD-experts
[23]	Children with ASD	Practical skill (healthcare)	3-walled CAVE system	–	–
[25]	People with NDD	Cognitive skill (attention)	Smartphone IVR headset (Google Cardboard)	Co-design	Special Educators Therapists Neuropsychiatric doctor
[29]	Adolescents with ASD	Social skill (interaction)	Tethered IVR headset (Oculus Rift)	–	–
[16,34]	Children with ASD	Practical skill (psycho-motor)	Large-scale screen projections	–	–
[39,40]	Children with ASD	Social skill (interaction)	Large-scale circular floor projections	User-Centered Design Participatory Design	Children with ASD
[41]	Children with ASD	Social Skill (Communication and social attention)	Tethered IVR headset (Oculus Rift)	–	–
[45]	Children with ASD	Cognitive skill (imitation and joint attention)	Tethered IVR headset (Oculus Rift)	–	–
[46]	Adults with HFA	Social skill (public speaking)	Smartphone IVR headset (Samsung Gear VR)	User-Centered Design Participatory Design	ICT experts Psychologists People with Asperger Digital artists
[47]	Children with HFA	Social and Cognitive skills	Tethered IVR headset (HCT Vive)	–	–
[48]	Children with ASD	Practical skill (public transportation)	Tethered IVR headset (Oculus Rift)	Agile Methodology	ASD-experts
[50]	Adolescents with ASD	Practical skill (work-related)	Large-scale circular floor projection	Emphatic Design	Teachers Students with ASD
[51]	Children with ASD	Social skill (interaction)	Tethered IVR headset (HCT Vive)	Human-Centered Design	–
[52]	Children with ASD	Social skill (emotion recognition)	3-walled CAVE system	–	Special Educators Therapists
[56]	People with NDD	Social skill (communication)	Smartphone IVR headset (Google Cardboard)	Co-design	Engineers NDD-experts
[58]	Children with ASD	Conceptual skill (language and literacy)	Smartphone IVR headset (3D glass)	Participatory Design	ASD-experts

Focusing on the retrieved HMD-based applications targeting children with ASD (40%), it is not clear if all the authors considered the scientific evidence regarding the recommended age for the safe use of HMDs (>13 years old) [21]: they mainly used the generic term "children" without specifying the age range. However, if we consider the age range of evaluation participants, in some cases, this advice was not respected (e.g., [41] targets children with ASD, but it was evaluated by children with ASD aged 6–10).

Learning Purpose. The main goal of retrieved IVR-based SG for people with ASD is to help them overcome the impairments implied by their clinical condition, especially social and communication ones. According to the classification provided by AAIDD [55], the included studies were categorized in conceptual skills, social skills, practical skills, and cognitive skills. Most of the included studies belong to the social skills category (45%) and, more specifically, they address interpersonal skills. For example, interaction skills [29,39,51], communication skills [41,56], emotion recognition skills [52], eye contact skills [22], public speaking skills [46], and social attention skills [41]. Among them, a particularly interesting study is presented by [39]. Indeed, players must collaboratively play each other to achieve the goal of the game (i.e., discover and catch different creatures hidden in the land of fog); thus, the targeted social skills are improved transparently and not directly through the game. In the practical skills category, we identified seven studies that aim to help people with ASD become more valuable to the society they live in. In fact, these studies promote the training of skills that could promote their independence and autonomy, such as the use of public transportation [48], street crossing [5], managing money [3] as well as healthcare [23], psycho-motor [34] and work-related skills [12,50]. In addition, one study belongs to the conceptual skill category, addressing language and literacy [58]. Finally, three studies were identified in the cognitive skill category, as they aim to train attention skills [25], imitation and joint attention skills [45], and non-verbal fluid intelligence, attention processes, and sensorimotor integration [20]. However, there was a study that cannot be associated with a single category: the IVR-based SG for people with ASD presented in [47] addressed skills belonging both to social and cognitive categories.

Technology. From the analysis of the included studies, it emerged that most of the SGs were administered as wearable IVR applications (65%). In detail, seven studies adopted tethered IVR headsets (i.e., HMD requiring a physical connection to a computer by cables) whereas five studies adopted smartphone IVR headsets (i.e., IVR systems that make use of smartphones to provide the IVR experience). Both solutions have pros and cons, especially in terms of quality of the IVR experience, system cost, and freedom of movement. On the one hand, tethered IVR headsets are much more immersive than other types of HMD due to the high-quality experience they can deliver [7]. In spite of these benefits, the cost of these systems is higher (since it includes not just the HMD but also the high-performance computer that runs the application), and the freedom of movement is restricted by cables. Moreover, these systems require a specified room space indicated by the guidelines of the producing company. On the other hand, smartphone IVR headsets provide limited immersion [7]. Indeed, the quality of the experience is related to the smartphone being used and appears, in any case, lower compared to that realised through tethered IVR headsets in terms of rendered graphics and refresh rates. However, these headsets are wireless and do not require a large room to be used. In addition, these are definitely cheaper compared to tethered IVR headsets as their cost ranged from few (e.g., Google Cardboard) to around a hundred euros (e.g., Samsung Gear VR) [7]. Regarding the identified tethered IVR headsets, four included studies used the HTC VIVE headset [3,22,47,51], while the others adopted the Oculus

Rift headset [29,41,45,48]. Instead, with respect to smartphone IVR headsets, it is possible to find Google Cardboard headsets [25,56], 3D glass [58], Xiaomi VR headset [5], and Samsung Gear VR headset [46].

Regarding the included studies reporting interventions delivered through large-scale projection-based systems, the following systems were found: 3-walled CAVE systems [23,52], large-scale screen projections [20,34], and large-scale circular floor projections [39,50]. All interventions required a very large space and different projectors, cameras, and sensors other than a computer with very high computational power.

In addition, a study providing an IVR-based SG for people with ASD articulated as a set of six mini-games was included: three delivered through an HMD (VR2200 headset), while the others through a large 180° curved curtain screen [12].

Design Methodology and Design Team. From the analysis of the studies retrieved, it emerged that over half of them (60% of included studies) reported relevant information regarding the design methodology followed and the composition of the team involved in the design itself.

Regarding the methodology followed to design the available IVR-based SGs, 50% of the retrieved studies include this information; however, it cannot completely rule out the possibility that others also adopted one. It emerged that the design methodologies extracted have the final users and their needs as the basis for the design, apart from one that followed the Agile methodology [48]. Specifically, these are Participatory Design/Co-Design [22,25,56,58], User-Centered Design [3], Human-Centered Design [51], and Emphatic Design [50]. In a few cases (10% of included studies), these are adopted jointly, i.e., User-Centered Design with Participatory design [39,46]. These findings are in agreement with the actual trend in ICT targeting people with ASD development [36,53]. In addition, the prevalence of Participatory Design is not surprising since it confirms the growing use of this design methodology in the development processes of ICT for people with ASD and the precious value of their contributions to design effective, accessible, and usable solutions [36].

Beyond the design methodology adopted, two studies specified the entire composition of the multidisciplinary team involved in the development process [46,56] while nine declared the participation of specific people, such as teachers [5,50], representatives of the target audience and/or their relatives [5,39,50,52,58], ASD experts [22,48,58], special educators [3,12,25,52], therapists [25,52], and neuropsychiatric doctor [25].

Focusing on how these individuals were involved in the development process, two studies specified the methods and approaches used to involve professional stakeholders and relatives of people with ASD, i.e., through unstructured focus group interviews [3] and structured questionnaires and semi-structured interviews [5]. Unfortunately, none of the studies described in detail how people with ASD were actually involved in the development process of IVR-based SG for such a population, besides reporting the design methodology adopted. This aspect is particularly crucial since activities to engage people with ASD should be carefully designed, given the unique characteristics of this clinical population.

5 Conclusion and Future Work

This review identified and analyzed 20 studies presenting IVR-based SGs for people with ASD. The results of this analysis presented an overview of the available applications and how previous research experience addressed the methodological and technological issues encountered.

The survey analysis has resulted in the following conclusions that can be considered as preliminary indications to develop IVR-based SGs for individuals with ASD.

- The launch of new high-quality and affordable IVR technologies has, since 2013, increased the number of publications on this topic. A progressive increase in such publications can still be expected.
- The available IVR-based SGs mainly target children with ASD. The adoption of such applications in the treatment of low functioning people with ASD and adults with ASD is still limited, although there is no empirical evidence that discouraged it. Although there are no recommendations on the age range for large-scale projection-based IVR technologies, the use of current HMDs is recommended for individuals older than 13 years [21].
- The adoption of tethered IVR headsets is widely common, due to the high-quality immersive experience they can deliver. However, this may hinder the large-scale deployment of these applications due to the high costs of the systems compared to more affordable IVR headset smartphones. However, in recent years, another type of IVR headset is being released that seems to be a good compromise, i.e., standalone IVR headsets (all-in-one) [7]; they provide a good quality of IVR experience, and their costs are more affordable compared to the thousand euros required by most powerful tethered IVR headsets.
- Available IVR-based SGs were mainly focused on social and practical skills learning. However, since IVR-based SGs are centered on visual representations - in line with the sensory preferences and visuospatial strength of people with ASD [29] - these may be seen as effective training interventions for the other skill categories as well.
- Although the design methodology is presented in a limited number of studies, it was observed that they were mainly developed following user-focused design methodologies enhancing the synergistic collaboration between different experts (e.g., in the fields of ICT, psychology, and ASD), such as Participatory Design and User-Centered Design.
- Despite the proven relevance [36], few studies declared an active participation of people with ASD in the design process; however, they did not specify how these participatory activities were carried out. In this regard, there is a

lack in the literature on holistic guidelines to translate traditional participatory design methods for typical development people into actual activities for effectively engaging ASD subjects [8,24]. Thus, further research is needed to identify the most appropriate way to ensure the effective engagement of this population in the development process [8,24,36].

Although IVR-based SG represents an innovative way to perform many learning and training tasks in ASD treatment, a lot of research work remains to be done to be able to cope with the challenges encountered since the design phase. Specifically, more comparison studies and reliable data are needed to identify benefits of different design choices, and leverage the future studies and systems.

The literature review and preliminary guidelines presented in this paper are meant to provide insight and motivation to future studies for investigating how to design, implement, and evaluate effective IVR-based SGs and, in the long run, leading to propose an innovative design methodology.

References

1. Abras, C., Maloney-Krichmar, D., Preece, J., et al.: User-centered design. In: Bainbridge, W. (ed.) Encyclopedia of Human-Computer Interaction, vol. 37, no. 4, pp. 445–456. Sage Publications, Thousand Oaks (2004)
2. Abt, C.C.: Serious Games. University Press of America (1987)
3. Adjorlu, A., Serafin, S.: Head-mounted display-based virtual reality as a tool to teach money skills to adolescents diagnosed with autism spectrum disorder. In: Brooks, A.L., Brooks, E., Sylla, C. (eds.) ArtsIT/DLI -2018. LNICST, vol. 265, pp. 450–461. Springer, Cham (2019). https://doi.org/10.1007/978-3-030-06134-0_48
4. Albrecht, M.A., Foster, J.K., Joosten, A., et al.: Visual search strategies during facial recognition in children with ASD. Res. Autism Spectr. Disord. **8**(5), 559–569 (2014)
5. Alharbi, A., Aloufi, S., Assar, R., Meccawy, M.: Virtual reality street-crossing training for children with autism in Arabic language. In: Professional International Conference on Innovation and Intelligence for Informatics, Computing and Technologies, pp. 1–6 (2020)
6. American Psychiatric Association: Diagnostic and Statistical Manual of Mental Disorders: DSM-5, vol. 5. American Psychiatric Association (2013)
7. Aniwaa Team: Types of VR headsets: PC VR, standalone VR, smartphone VR. https://www.aniwaa.com/guide/vr-ar/types-of-vr-headsets/. Accessed 01 Mar 2022
8. Benton, L., Vasalou, A., Khaled, R., et al.: Diversity for design: a framework for involving neurodiverse children in the technology design process. In: Proceedings of the SIGCHI Conference on Human Factors in Computing Systems, pp. 3747–3756. ACM (2014)
9. Bernardini, S., Porayska-Pomsta, K., Smith, T.J.: ECHOES: an intelligent serious game for fostering social communication in children with autism. Inf. Sci. **264**, 41–60 (2014)
10. Boucenna, S., Narzisi, A., Tilmont, E., et al.: Interactive technologies for autistic children: a review. Cogn. Comput. **6**(4), 722–740 (2014). https://doi.org/10.1007/s12559-014-9276-x

11. Bozgeyikli, E., Bozgeyikli, L., Raij, A., Katkoori, S., Alqasemi, R., Dubey, R.: Virtual reality interaction techniques for individuals with autism spectrum disorder: design considerations and preliminary results. In: Kurosu, M. (ed.) HCI 2016. LNCS, vol. 9732, pp. 127–137. Springer, Cham (2016). https://doi.org/10.1007/978-3-319-39516-6_12
12. Bozgeyikli, L., et al.: VR4VR: vocational rehabilitation of individuals with disabilities in immersive virtual reality environments. In: Proceedings of the 8th ACM International Conference on PErvasive Technologies Related to Assistive Environments, pp. 1–4 (2015)
13. Bozgeyikli, L., Bozgeyikli, E., Raij, A., Alqasemi, R., Katkoori, S., Dubey, R.: Vocational rehabilitation of individuals with autism spectrum disorder with virtual reality. ACM Trans. Access. Comput. **10**(2), 1–25 (2017)
14. Bozgeyikli, L., Bozgeyikli, E., Raij, A., et al.: Vocational training with immersive virtual reality for individuals with autism: towards better design practices. In: IEEE 2nd Workshop on Everyday Virtual Reality, pp. 21–25 (2016)
15. Bozgeyikli, L., Raij, A., Katkoori, S., Alqasemi, R.: A survey on virtual reality for individuals with autism spectrum disorder: design considerations. IEEE Trans. Learn. Technol. **11**(2), 133–151 (2018)
16. Cai, Y., Chia, N.K., Thalmann, D., Kee, N.K., Zheng, J., Thalmann, N.M.: Design and development of a virtual dolphinarium for children with autism. IEEE Trans. Neural Syst. Rehabil. Eng. **21**(2), 208–217 (2013)
17. Camargo, M.C., Barros, R.M., Brancher, J.D., Barros, V.T.O., Santana, M.: Designing gamified interventions for autism spectrum disorder: a systematic review. In: van der Spek, E., Göbel, S., Do, E.Y.-L., Clua, E., Baalsrud Hauge, J. (eds.) ICEC-JCSG 2019. LNCS, vol. 11863, pp. 341–352. Springer, Cham (2019). https://doi.org/10.1007/978-3-030-34644-7_28
18. Caruso, F., Di Mascio, T.: Designing IVR serious games for people with ASD. In: Kubincová, Z., Lancia, L., Popescu, E., Nakayama, M., Scarano, V., Gil, A.B. (eds.) MIS4TEL 2020. AISC, vol. 1236, pp. 291–295. Springer, Cham (2021). https://doi.org/10.1007/978-3-030-52287-2_30
19. Checa, D., Bustillo, A.: A review of immersive virtual reality serious games to enhance learning and training. Multimed. Tools. Appl. **79**(9), 5501–5527 (2020). https://doi.org/10.1007/s11042-019-08348-9
20. De Luca, R., Naro, A., Colucci, P.V., Pranio, F., et al.: Improvement of brain functional connectivity in autism spectrum disorder: an exploratory study on the potential use of virtual reality. J. Neural Transm. **128**(3), 371–380 (2021). https://doi.org/10.1007/s00702-021-02321-3
21. Di Mascio, T., Tarantino, L., De Gasperis, G., Pino, C.: Immersive virtual environments: a comparison of mixed reality and virtual reality headsets for ASD treatment. In: Gennari, R., et al. (eds.) MIS4TEL 2019. AISC, vol. 1007, pp. 153–163. Springer, Cham (2020). https://doi.org/10.1007/978-3-030-23990-9_19
22. Elgarf, M., Abdennadher, S., Elshahawy, M.: I-interact: a virtual reality serious game for eye contact improvement for children with social impairment. In: Alcañiz, M., Göbel, S., Ma, M., Fradinho Oliveira, M., Baalsrud Hauge, J., Marsh, T. (eds.) JCSG 2017. LNCS, vol. 10622, pp. 146–157. Springer, Cham (2017). https://doi.org/10.1007/978-3-319-70111-0_14
23. Finkelstein, S., Barnes, T., Wartell, Z., Suma, E.A.: Evaluation of the exertion and motivation factors of a virtual reality exercise game for children with autism. In: 1st Workshop on Virtual and Augmented Assistive Technology (VAAT), pp. 11–16. IEEE (2013)

24. Frauenberger, C., Makhaeva, J., Spiel, K.: Blending methods: developing participatory design sessions for autistic children. In: Proceedings of the International Conference on Interaction Design and Children, pp. 39–49 (2017)
25. Gelsomini, M.: An affordable virtual reality learning framework for children with neuro-developmental disorder. In: Proceedings of the 18th International ACM SIGACCESS Conference on Computers and Accessibility, pp. 343–344. Association for Computing Machinery, New York (2016)
26. Glaser, N., Schmidt, M.: Systematic literature review of virtual reality intervention design patterns for individuals with autism spectrum disorders. Int. J. Hum.-Comput. Int. **38**(8), 753–788 (2021)
27. Grossard, C., Grynspan, O., Serret, S., et al.: Serious games to teach social interactions and emotions to individuals with autism spectrum disorders (ASD). Comput. Educ. **113**, 195–211 (2017)
28. Hassan, A., Pinkwart, N., Shafi, M.: Serious games to improve social and emotional intelligence in children with autism. Entertain. Comput. **38**, 100417 (2021)
29. Herrero, J.F., Lorenzo, G.: An immersive virtual reality educational intervention on people with autism spectrum disorders (ASD) for the development of communication skills and problem solving. Educ. Inf. Technol. **25**(3), 1689–1722 (2019). https://doi.org/10.1007/s10639-019-10050-0
30. Khowaja, K., Salim, S.S., Al-Thani, D.: Components to design serious games for children with autism spectrum disorder (ASD) to learn vocabulary. In: Proceedings of the 5th International Conference on Engineering Technologies and Applied Sciences, pp. 1–7. IEEE (2018)
31. Kitchenham, B.: Procedures for performing systematic reviews **33**(2004), 1–26. Keele University, Keele, UK (2004)
32. Lai, M.C., Lombardo, M.V., Baron-Cohen, S.: Autism Lancet **383**(9920), 896–910 (2014)
33. Lorenzo, G., Lledó, A., Arráez-Vera, G., Lorenzo-Lledó, A.: The application of immersive virtual reality for students with ASD: a review between 1990–2017. Educ. Inf. Technol. **24**(1), 127–151 (2019). https://doi.org/10.1007/s10639-018-9766-7
34. Lu, A., Chan, S., Cai, Y., et al.: Learning through VR gaming with virtual pink dolphins for children with ASD. Interact. Learn. Environ. **26**(6), 718–729 (2018)
35. Marne, B., Wisdom, J., Huynh-Kim-Bang, B., Labat, J.-M.: The six facets of serious game design: a methodology enhanced by our design pattern library. In: Ravenscroft, A., Lindstaedt, S., Kloos, C.D., Hernández-Leo, D. (eds.) EC-TEL 2012. LNCS, vol. 7563, pp. 208–221. Springer, Heidelberg (2012). https://doi.org/10.1007/978-3-642-33263-0_17
36. Maun, R., Fabri, M., Trevorrow, P.: Adapting participatory design activities for autistic adults: a review. In: Soares, M.M., Rosenzweig, E., Marcus, A. (eds.) HCII 2021. LNCS, vol. 12780, pp. 300–314. Springer, Cham (2021). https://doi.org/10.1007/978-3-030-78224-5_21
37. Mesa-Gresa, P., Gil-Gómez, H., Lozano-Quilis, J.A., Gil-Gómez, J.A.: Effectiveness of virtual reality for children and adolescents with autism spectrum disorder: an evidence-based systematic review. Sensors **18**(8), 2486 (2018)
38. Miller, H.L., Bugnariu, N.L.: Level of immersion in virtual environments impacts the ability to assess and teach social skills in autism spectrum disorder. Cyberpsychology Behav. Soc. Netw. **19**(4), 246–256 (2016)
39. Mora-Guiard, J., Crowell, C., Pares, N., Heaton, P.: Lands of fog: helping children with autism in social interaction through a full-body interactive experience.

In: Proceedings of the 15th International Conference on Interaction Design and Children, pp. 262–274 (2016)
40. Mora-Guiard, J., Crowell, C., Pares, N., Heaton, P.: Sparking social initiation behaviors in children with Autism through full-body interaction. Int. J. Child-Comput. Interact. **11**, 62–71 (2017)
41. Naranjo, C.A., Ortiz, J.S., Álvarez, V.M., et al.: Teaching process for children with autism in virtual reality environments. In: Proceedings of the 9th International Conference on Education Technology and Computers, pp. 41–45 (2017)
42. Park, J.H., Abirached, B., Zhang, Y.: A framework for designing assistive technologies for teaching children with ASDs emotions. In: CHI 2012 Extended Abstracts on Human Factors in Computing Systems, pp. 2423–2428. ACM (2012)
43. Parsons, S.: Authenticity in virtual reality for assessment and intervention in autism: a conceptual review. Educ. Res. Rev. **19**, 138–157 (2016)
44. Peretti, S., Mariano, M., Mazzocchetti, C., et al.: Diet: the keystone of autism spectrum disorder? Nutr. Neurosci. **22**(12), 825–839 (2019)
45. Rahmadiva, M., Arifin, A., Fatoni, M.H., et al.: A design of multipurpose virtual reality game for children with autism spectrum disorder. In: Proceedings International Conference on Biomedical Instrumentation and Technology, vol. 1, pp. 1–6. IEEE (2019)
46. Rojo, D., Mayor, J., Rueda, J.J.G., Raya, L.: A virtual reality training application for adults with Asperger's syndrome. IEEE Comput. Graph. Appl. **39**(2), 104–111 (2019)
47. Shahab, M., Taheri, A., Mokhtari, M., et al.: Utilizing social virtual reality robot (V2R) for music education to children with high-functioning autism. Educ. Inf. Technol. **27**(1), 819–843 (2021). https://doi.org/10.1007/s10639-020-10392-0
48. Simões, M., Bernardes, M., Barros, F., et al.: Virtual travel training for autism spectrum disorder: proof-of-concept interventional study. JMIR Serious Games **6**(1), e8428 (2018)
49. Soloway, E., Guzdial, M., Hay, K.E.: Learner-centered design: the challenge for HCI in the 21st century. Interactions **1**(2), 36–48 (1994)
50. Takahashi, I., Oki, M., Bourreau, B., Suzuki, K.: Designing interactive visual supports for children with special needs in a school setting. In: Proceedings of the International Conference on Designing Interactive Systems, pp. 265–275 (2018)
51. Thordarson, A., Vilhjálmsson, H.H.: SoCueVR: virtual reality game for social cue detection training. In: Proceedings of the ACM International Conference on Intelligent Virtual Agents, pp. 46–48 (2019)
52. Tsai, W.T., Lee, I.J., Chen, C.H.: Inclusion of third-person perspective in CAVE-like immersive 3D virtual reality role-playing games for social reciprocity training of children with an autism spectrum disorder. Univ. Access Inf. Soc. **20**(2), 375–389 (2020). https://doi.org/10.1007/s10209-020-00724-9
53. Tsikinas, S., Xinogalos, S.: Studying the effects of computer serious games on people with intellectual disabilities or autism spectrum disorder: a systematic literature review. J. Comput. Assist. Learn. **35**(1), 61–73 (2019)
54. Valencia, K., Rusu, C., Quiñones, D., Jamet, E.: The impact of technology on people with autism spectrum disorder: a systematic literature review. Sensors **19**(20), 4485 (2019)
55. Vannest, K.J.: AAIDD (American Association on Intellectual Developmental Disabilities) Classification Systems. Encyclopedia of Special Education: A Reference for the Education of Children, Adolescents, and Adults with Disabilities and Other Exceptional Individuals (2013)

56. Vona, F., Silleresi, S., Beccaluva, E., Garzotto, F.: Social MatchUP: collaborative games in wearable virtual reality for persons with neurodevelopmental disorders. In: Ma, M., Fletcher, B., Göbel, S., Baalsrud Hauge, J., Marsh, T. (eds.) JCSG 2020. LNCS, vol. 12434, pp. 49–65. Springer, Cham (2020). https://doi.org/10.1007/978-3-030-61814-8_4
57. Whyte, E.M., Smyth, J.M., Scherf, K.S.: Designing serious game interventions for individuals with autism. J. Autism Dev. Disord. **45**(12), 3820–3831 (2015). https://doi.org/10.1007/s10803-014-2333-1
58. Winoto, P., Xu, C.N., Zhu, A.A.: "Look to remove": a virtual reality application on word learning for Chinese children with autism. In: Antona, M., Stephanidis, C. (eds.) UAHCI 2016. LNCS, vol. 9739, pp. 257–264. Springer, Cham (2016). https://doi.org/10.1007/978-3-319-40238-3_25
59. Zakari, H.M., Ma, M., Simmons, D.: A review of serious games for children with autism spectrum disorders (ASD). In: Ma, M., Oliveira, M.F., Baalsrud Hauge, J. (eds.) SGDA 2014. LNCS, vol. 8778, pp. 93–106. Springer, Cham (2014). https://doi.org/10.1007/978-3-319-11623-5_9
60. Zhao, Q.: A survey on virtual reality. Sci. China Inf. Sci. **52**(3), 348–400 (2009). https://doi.org/10.1007/s11432-009-0066-0

Author Index

M. Temperini et al. (Eds.): MIS4TEL 2022, LNNS 580, pp. 197–198, 2023.
https://doi.org/10.1007/978-3-031-20617-7